GENIUS

of BRITAIN

Petrus Alfonsi
1062–c1130

Adelard of Bath
c1080–c1152

Bartholomew the Englishman
c1203–1272

ert Grosseteste
c1170–1253

Roger Bacon
c1214–c1294

Francis Bacon
1561–1626

William Harvey
1578–1657

pier
17

William Oughtred
1575–1660

Robert Hooke
1635–1703

Thomas Savery
c1650–1715

Isaac Newton
1643–1727

John Hunter
1728–1793

James Hutton
1726–1797

Joseph Black
1728–1799

Daniel Rutherford
1749–1819

William Murdock
c1754–1839

GENIUS
of BRITAIN
The Scientists who Changed the World

ROBERT UHLIG

FOREWORD BY JAMES DYSON

Collins

To Hilary, Linus and Truman

First published by Collins 2010

HarperCollins*Publishers*
77–85 Fulham Palace Road,
Hammersmith, London W6 8JB

www.harpercollins.co.uk

1 3 5 7 9 10 8 6 4 2

© Robert Uhlig 2010

Robert Uhlig asserts the moral right to
be identified as the author of this work

A catalogue record of this book
is available from the British Library

ISBN 978 0 00 732067 7

Printed and bound in Great Britain by
Clays Ltd, St Ives plc

Mixed Sources
Product group from well-managed
forests and other controlled sources
www.fsc.org Cert no. SW-COC-001806
© 1996 Forest Stewardship Council

FSC is a non-profit international organisation established to promote the
responsible management of the world's forests. Products carrying the FSC
label are independently certified to assure consumers that they come
from forests that are managed to meet the social, economic and
ecological needs of present and future generations.

Find out more about HarperCollins and the environment at
www.harpercollins.co.uk/green

CONTENTS

ACKNOWLEDGEMENTS

This book was written to accompany the Channel 4 series *Genius of Britain*. Although some of the content was drawn from the television series, the book spans a wider period and includes the achievements and lives of many more great British scientists than broadcast, as well as several émigré scientists who made their most significant discoveries in Britain.

Many thanks to Professor Jim Al-Khalili, Sir David Attenborough, Richard Dawkins, Sir James Dyson, Professor Stephen Hawking, Professor Sir Paul Nurse, Professor Kathy Sykes and Professor Lord Robert Winston for allowing extracts from their commentary in *Genius of Britain* to be used in this book. Most of the comments are drawn directly from the production of the series. In some cases, they have been edited by the presenter to suit the printed medium. For Stephen Hawking, it is extremely important to emphasise that his comments are exactly as spoken in the series.

Copious thanks are owed to everyone at RDF and IWC Media who assisted with the genesis of this book, in particular Mark Lesbirel, Victoria Ramirez, Peter Collins and Rachel Bell. Thanks also to Susanna Abbott, Denise Bates, Hannah MacDonald and Helena Nicholls at HarperCollins. For excellent guidance and advice, many thanks to Julian Alexander and everyone at LAW.

PICTURE CREDITS

FOREWORD

It's one of the fundamental things that makes us human: wondering why the world is the way it is. For some of us, it is enough merely to wonder. For most, basic explanations of why and how we came to be here satisfy a casual curiosity. But for a special few – the British geniuses featured in this book – entire lives are dominated by posing questions that no one has asked before, and then finding the answers.

That's what this book is about: the lives and achievements of the Britons who discovered and decoded the mysteries of the universe. Men and women who changed our perception of ourselves and of our surroundings from a belief in mystical superstitions to rational understandings of our existence. Household names such as Isaac Newton, Charles Darwin and Michael Faraday, but also lesser-known geniuses such as J.J. Thompson, John Hunter and Fred Hoyle.

This history of British science and its scientists begins in the late seventh century, when Vikings were overrunning the last vestiges of Roman culture. Only a few monks in the north-east of England were keeping scientific enquiry alive by studying and translating classical Greek and Roman philosophical works of nature, medicine, astronomy and arithmetic. But in this tiny pocket of philosophical learning the course was being set for the unimaginably rich and fascinating journeys of scientific exploration that continue to this day.

And what journeys they were! Over the next thirteen centuries British scientists would embark on great sagas of endurance involving determination, cunning, rivalry and co-operation. These are men and women who through blood, sweat and tears overcame all the obstacles to find out the truth. Theirs are stories of obsession, of people who pushed the boundaries and who injected themselves with syphilis or drank poison or blew themselves up or undertook suicidal voyages – all in the name of discovery. And as successive generations of British scientists built upon the work of their predecessors in their quest for truth and understanding, they were beset by frustrations, setbacks and disasters as frequently as triumphs, serendipities and celebrations. But ultimately their quest for knowledge and understanding would change the course of history and shape civilisation.

We Britons have so much of which to be proud with respect to our role in the history of science. As a nation, we punch a long way above our weight – perhaps more so over the last 400 years than any other nation. And yet I sometimes wonder whether we really appreciate our scientific heritage. Most of us know that Britain was the cradle of the Industrial Revolution, but how many of us realise that it was also the birthplace of the Scientific Revolution? And that many of the world's greatest scientists were either Britons or did their greatest work in Britain?

We may only be a small island, but we are far from small minded. Britain's great inventors and scientists have taken mankind to places and into worlds that we only dreamed possible. The steam engine, evolution, the atom, the computer and the world wide web – British science has quite literally created the modern world. And in this book, and the Channel 4 television series that it accompanies, the stories of the people who did it are told.

Albert Einstein famously kept three photographs on the wall of his study of the three men he considered the true giants of science, and all of them were Britons: Isaac Newton, Michael Faraday and James Clerk Maxwell. To this illustrious company, add Charles Darwin, or the great experimentalists Robert Hooke and Robert Boyle, or John Hunter, the pioneer of modern surgery, and Edward Jenner, the man who took enormous risks to prove the principle of vaccination, and then there's Alexander Fleming, who discovered penicillin. The list goes on. Thomas Savery and James Watt (and a supporting cast that included Richard Trevithick and William Murdock) fathered the steam engine, one of the greatest and most influential of inventions. Beginning with

FOREWORD

J.J. Thompson, two generations of great British minds made most of the fundamental discoveries concerning atomic and nuclear structure, including various forms of nuclear radiation and the first splitting of the nucleus. And it's not over yet. In Cambridge, Professor Alan Windle is creating the wonder material of the future, carbon nanotubes.

Thanks to these great Britons, Britain has played a leading role in the discoveries that have helped us understand our place in the universe. It happened because these scientists all shared a curiosity about the world – about how the universe works and why.

Curiosity about a falling apple led Newton to wonder about the fundamental forces that control everything in the universe. It led Newton to unlock the secrets of the physical world and to discover the sublime beauty of gravity. But it also led Newton to become one of the most secretive and vindictive of men, whose obsessions would drive him to the edge of insanity.

Curiosity about subtle but highly specialised differences between species led Darwin to wonder how and why all the species originated. Thereby he discovered evolution by natural selection and unleashed a theory that has consumed our culture and questioned the existence of God. Curiosity about why a wire carrying a current from a battery generated a magnetic field led Faraday to discover the electromagnetic relationship and to develop the electric motor and the dynamo, without which the modern world would not exist. And a curiosity about what happened at the beginning of time led Fred Hoyle to work out the formation of all the elements in the universe – the building blocks of quite literally everything.

Even the fate of nations is determined by our scientific curiosity. It shaped our past and the world in which we live. And now more than ever, science will decide our future. Our prosperity, happiness and success will depend on our ability to understand complex scientific issues and to develop new ways of dealing with them.

Science is of such fundamental importance because it gives us a systematic framework for questioning what is happening in the world. Climate change; new drugs for cancer and other life-threatening illnesses; the internet; test-tube babies; space travel; the atom smasher at CERN; the threat of nuclear weapons and bioterrorism; food shortages and over-population; the use of vaccination to combat virus pandemics – all of these require an appreciation of science and the scientific method if they are to be understood. Consequently, it is crucial for democracy that the general public appreciates science and how it works.

Fortunately, we are all born curious. From our youngest days we want to understand what's going on around us. I suppose you could say this inherent curiousness means we are all born scientists, or at least with a scientist's attitude of intelligent open-minded enquiry.

But for some reason, as we get older many of us lose our curiosity about the functioning of the world. Some of us change from wide-eyed toddlers into disinterested cynics. We stop wondering why and how things work. We favour easily understood but untested and frequently mystical explanations of the world instead of rational and empirical descriptions. That's fatal.

The great scientists in this book never lost their fascination for understanding, knowledge and truth. And if we want to participate in the great human journey we need to make sure that future generations never lose their quest and desire to ask fundamental and probing questions about our existence. We have to ensure that science education is a priority and that our great universities can maintain their science departments.

The first step to ensuring future generations will be informed participants in our scientific future is an appreciation of our scientific past. I hope that this book and the stories it tells of great British geniuses and their achievements will inspire you and your children to learn more about our great scientific heritage.

And if we are enthused about the fantastic lives and achievements of our greatest scientific geniuses, hopefully we will be motivated to follow in our predecessors' illustrious footsteps. We know we'll need highly trained scientists and engineers for the challenges that we face – challenges such as developing non-fossil energy sources or ways of guaranteeing future food supplies or nanotechnologies to provide new materials and machines. We'll also need politicians, bureaucrats and people in all walks of life who understand the scientific principles behind the issues we'll face. I hope this book will inspire and help everyone who reads it to take a small step along that path of enlightenment and to keep on walking to its fascinating end.

James Dyson
January 2010

Chapter 1

IDEAS FROM ABROAD

Bread Street in the City of London is an unlikely setting for the birthplace of modern science. Nowadays it's little more than a dusty canyon squeezed between a succession of monolithic grey office blocks. There's a sandwich bar, a few delivery bays and, at its junction with Watling Street, a tantalising glimpse of the magnificent baroque dome of St Paul's Cathedral. Built by Sir Christopher Wren after the Great Fire of London in 1666, the cathedral is a visual reminder of why, as a result of the fire and the 1940 'Blitz' bombardment of London, nothing remains to remind us of Bread Street's remarkable past, when it teemed with intellect, enquiry and creativity. Where Bread Street meets Friday Street, the most notable writers and thinkers of the day – including Shakespeare, Ben Jonson and John Donne, who like his fellow poet, John Milton, was born in Bread Street – would meet to battle wits in a club founded at the Mermaid Tavern by Sir Walter Raleigh. Meanwhile, in the nearby churchyard of the old gothic St Paul's, booksellers would set up their stalls while choirboys (like the 'little eyases' that Hamlet complained competed with his players) performed scenes from the latest plays. However, the location that can lay claim to playing a starring role in the opening act of the scientific revolution was further down Bread Street towards the river, where Peter Short, a freeman of the Stationers Company, operated a successful printing business behind a handsome shop front under a star-shaped sign.

In early 1600 Short was busy with books for Jonson and Marlowe. Having printed several of Shakespeare's first editions and early texts in the previous three years, Short was in high demand, but the most influential book he would print that year – or any year, for that matter – belonged to William Gilbert, personal physician to Elizabeth I and president of the College of Physicians. If any one work marked the start of the scientific revolution – that moment when rational, empirical, experimental investigation replaced mysticism, conjecture and superstition as the means of explaining the world – then Gilbert's book was it.

Concerned primarily with magnetism, electricity and astronomy, *De magnete, magneticisque corporibus, et de magno magnete tellure* ('*A new natural philosophy of the lodestone, magnetic bodies, and the great lodestone the earth, proved by many reasonings and experiments*') was the first work under the modern definition of physical science to be produced anywhere in the world. Its principal ideas were so remarkable that they would not be added to until Michael Faraday's discoveries some 230 years later. However, the fact that these ideas had been formulated by an eminent medic who devoted eighteen years of his spare time (and several million pounds in modern money) to the task was only half the story. The most remarkable aspect of *De magnete* was not the content or the new theories that Gilbert put forward, but the way in which he formulated them.

For nearly 2,000 years science had relied on the writings of natural philosophers and mathematicians such as Aristotle, Pliny and Copernicus, who had published treatises on astronomy, geometry and the motion of heavenly and terrestrial bodies. They'd discussed the anatomy of animals, the structure of plants and the classification of species. Alchemists among them had searched for ways to turn lead into gold and physicists had declared the world to be constructed of four fundamental elements – earth, water, air and fire – built up in consecutive shells.

Although the conclusions of these early scientists and their successors differed, they all relied on applying philosophical methods in their attempts to understand the world. Using logical discourse and scholastic interpretation of earlier texts to develop their theories, they had developed an academic tradition that Gilbert shattered with his publication of *De magnete*.

Gilbert's momentous breakthrough was to assert that nothing could be taken for granted or postulated if it could not be proved by extensive observations from repeatable experiments. This ethos, the bedrock of modern science, made Gilbert in effect the first scientist, although the term would not be coined for another 230 years. His book, a bestseller by the standards of its day (it was even pirated in counterfeit editions), was hugely influential and a profound influence on his contemporary Galileo Galilei, who is often regarded as the 'father of science' but who lauded Gilbert as the founder of the experimental method for which Galileo is usually given credit.

Like many iconoclasts, Gilbert travelled an unconventional route to his intellectual breakthrough. Having qualified at Cambridge University as a medic in 1569, he rapidly established himself as a physician to the aristocracy and court, which led to the Privy Council asking him to treat some sailors. This brought him into contact with Sir Francis Drake and his fellow Elizabethan circumnavigator, Thomas Cavendish.

Gilbert's respect for these heroic mariners appears to have triggered a fascination with the nautical compass and magnetism, perhaps unsurprising as the compass was the most significant invention of its day. By increasing the safety and scope of sea voyages, the compass had opened up the Eastern hemisphere to Western travellers and made possible the age of exploration. It had played no small role in the settlement of North America in the 1580s and the sea battles that led to the defeat of the Spanish Armada in 1588. It is safe to say that the compass's enabling role in trade, imperialism, warfare and missionary exploration had a greater influence on the course of history than the invention of either gunpowder or the printing press, and that it would push man's perception of the world further than even Copernican astronomical theories.

Gilbert brought a fresh pair of eyes to the many superstitions that surrounded the mysterious workings of the compass. His curiosity and scepticism were in keeping with his habit over the years of questioning the adherence by his colleagues to the texts of the Greek physician, Galen, which explained the workings of the body as a collection of four mystical humours – blood, black bile, yellow bile and phlegm – formed from the fundamental elements of fire, water, air and earth. According to Galenic medicine, disease resulted from the imbalance of these humours or the dominance of one of the four qualities of hot, cold, wet and dry.

Although Gilbert had been taught Galenic medicine at Cambridge and the Galenic method was promoted by the College of Physicians, he dismissed its orthodoxy as rooted essentially in abstract philosophy. Instead Gilbert advocated an experience-based medicine and applied the same empirical mindset to his examination of magnetism and the workings of the compass.

In his introduction to *De magnete*, Gilbert did not pull any punches in rejecting the orthodox natural philosophy of his day. Damning some long-held beliefs, he commented that 'in philosophy many false and idle conjectures arise from fables and falsehoods' – brave words at a time when philosophers were burned at the stake for the heresy of challenging established Church-approved doctrine in other European countries. The opening sentence of *De magnete*'s prologue castigated his predecessors, accusing them of promoting theories 'on the basis of a few vague and indecisive experiments'. Gilbert continued in a provocative and defiant dismissal of his predecessors' work, declaring that 'clearer proofs in the discovery of secrets and the investigations of the hidden causes of things are afforded by trustworthy experiments and by demonstrated arguments than by the probable guesses and opinions of the ordinary professors of philosophy'.

The book then described a series of brilliant and pioneering experiments, in the first tranche of which Gilbert demolished many common misnomers about magnets. At a time when sailors would be flogged for the offence of having garlic on their breath, Gilbert proved that it was impossible for garlic to demagnetise the ship's compass or any other magnet, or incidentally for it to cure headaches. He then went on to show that a compass needle points along a roughly north–south axis and that it dips downwards if it is suspended. By examining the degree of dip of a compass needle in the vicinity of a spherical magnet, he showed that the needle pointed vertically at the magnetic poles of the sphere, which led him to declare that the Earth itself acted like a giant bar magnet. To acknowledge the similarities between the Earth's magnetic field and that of a bar magnet, Gilbert was the first person to name the ends of a magnet its north and south poles.

Having overturned centuries of mysticism surrounding magnetism, Gilbert turned to other attractive forces. He discovered that amber, rock crystal and several gems would attract almost any light object when rubbed with silk. Realising that there was a distinction between this force and magnetism, which attracted only iron, Gilbert grouped all substances that showed the property under the name 'electrics', coining it from the Greek word for amber, *elektron*.

Gilbert's passion for enquiry was unstoppable. Having become the first to make the distinction between static electricity and magnetism, he then turned to the heavens. As the first notable British supporter of the Copernican view that the Sun was at the centre of the universe with the Earth orbiting around it – not vice versa, as advocated by the Church – he devised elegant explanations for several hitherto unexplained astronomical phenomena. He also speculated that magnetism kept Earth on its celestial track, a conclusion that would not be bettered by Galileo or Johann Kepler, the German astronomer whose laws of planetary motion would later provide the foundations for Newton's theory of universal gravitation. (Both Galileo and Kepler drew heavily on Gilbert in their promotion of Copernicus.)

Reading Gilbert's description of his experiments in *De magnete* gives a sense of a free-thinking innovator struggling at times to impose a logical pattern on frequently puzzling and contradictory observations. For instance, Gilbert noticed that magnetic forces persisted across a flame, but that magnetic iron lost its power when raised to red heat. He also discovered that water moisture in breath disrupted static electricity but a coating of oil did not, and that droplets of water were themselves attracted by electric forces.

Packed with such observations, *De magnete* paints a picture of science in its rawest state. Newton would write nearly ninety years later of standing 'on the shoulders of giants' when coming to his conclusions. Forced to formulate his understanding from the most basic principles and observations, Gilbert had no such luxury and, in so doing, became one of the giants to which Newton would later refer. Only three years after *De magnete* was published, Gilbert died, most probably in the plague epidemic of 1603 that also killed his printer, Peter Short, but his influence had already shaken the world of science, prompting Kepler to write that he wished he 'had wings with which to travel to England to confer with him'.

Although Gilbert undoubtedly drew on ideas coming out of the Renaissance in Italy, his fervent belief in experimentalism and his dismissal of the conventions of natural philosophy were all the more remarkable because they appeared at the time almost to come out of nowhere. For 250 years before Gilbert, scientific investigation had ground to a halt in Britain, largely as the result of disease and war. The great famine of 1315–17 and the Black Death, which entered England in 1348 through the port of Weymouth, killing up to half of the country's population by 1666, had predictably devastating effects. In the century from 1276 to 1375, average life expectancy more than halved from 35 to 17 years. Many of those who survived or were born after these two natural threats to life were dragged into the Hundred Years' War of 1337–1453 or the immediately succeeding Wars of the Roses of 1453–87. Unsurprisingly, two centuries of death and destruction revitalised interest in religion (and with it, a suspicion of non-religious explanations for the universe), while the sharp decline in available labour prompted draconian legislation that led to social unrest and a rise in criminality. It turned Britain into a place in which scientific investigation was low on the agenda.

But before the destructive events of the fourteenth and fifteenth centuries, developments in science – or more properly, natural philosophy, as it was called – had been advancing rapidly by Western European standards. The source of this rich heritage in scientific learning can be found another 500 years or so earlier in the late seventh century, when the last vestiges of Roman civilisation in Britain were being overrun by Viking invaders. Like most of Western Europe, Britain had become a tapestry of rural populations and semi-nomadic people since the political disintegration of Rome nearly 300 years earlier. Four frequently warring cultures – Celtic-speaking Romano-British in the west, pagan Picts in Scotland, the Dal Riata Gaels in Ireland, and Anglo-Saxons and Jutes along the east coast – were sharing a territory that under the Romans had been largely unified as Britannia. The downfall of urban life had reduced the scope of learning and the only remnants of scholarship were now found in places such as Lindisfarne monastery, where monks worked tirelessly, copying sacred and historic texts to ensure the survival of early Greek, Latin and Christian literature.

For many of these clergymen, the study of nature formed only a tiny part of their interest. With little institutional support for the study of natural phenomena, they concentrated their attention on religious topics. Nature was studied more for practical reasons than abstract enquiry. The need to care for the sick led to the study of medicine and of ancient texts on drugs. The quest for determining the proper time to pray led them to examine the motion of the stars. And a requirement to compute the date of Easter led them to explore and teach rudimentary mathematics and the motions of the Sun and Moon.

Among these monks was 'the Venerable' Bede, often called the father of English history for his most famous work, *The Ecclesiastical History of the English People*. Born on the lands of the Monastery of Saints Peter and Paul at Wearmouth and Jarrow in around 672, Bede was entrusted to the care of the monks at the age of 7. By the time he was ordained a deacon at 19 Bede was a conscientious choir attendee, but he refused higher office after he entered the priesthood, preferring to spend his time studying the writings of Greek and Roman philosophers, astronomers and mathematicians. Drawing on works by Aristotle, Pliny and Sosigenes held by the monastery library, this astonishingly versatile scholar had by 703 produced the first British work on science. *De temporibus* (*On Time*) was concerned mainly with calculating the date of Easter and became a standard text for the Church. It also included a new chronology of the world that placed the date of creation as 3952 BC, which had the effect of suggesting that Christ was not incarnated at the time advocated by the Church. Enraged that Bede had departed from the precise chronology of the Six Ages of the World theory accepted by theologians at the time, a group of drunken monks accused him of heresy at a dinner in front of Wilfred, Bishop of Hereford. Bede defended the accusation in a letter to Wilfrid, but didn't desist from continuing to challenge orthodox beliefs.

Twenty years later, in about 723, Bede wrote a longer work, a codex called *De temporium ratione* (*On the Reckoning of Time*). Many centuries before Renaissance scientists in Italy came to the same conclusions, it suggested that the world was round and that its spherical shape could explain the lengthening and shortening of daylight hours. With chapters on how the relative positions of the Sun and Moon influenced the appearance of New Moons at evening twilight, it also suggested how the Moon and latitude affected tidal cycles. Bede also highlighted shortcomings in the accuracy of the Julian calendar, warning that it would eventually put Easter out of phase with the March equinox and place the months out of synchrony with the seasons.

The Julian calendar had been introduced by Julius Caesar on the advice of Sosigenes, an Alexandrian astronomer. It advocated one leap year every four years to maintain synchrony with the solar cycle. But Bede warned that this adjustment was slightly inaccurate and that adhering to it would ultimately create chaos. In spite of Bede's warnings, it took more than 1,000 years for the error to be addressed in Britain. By then, it was necessary to correct by 11 days, so Wednesday 2 September 1752 was followed by Thursday 14 September 1752. Ever since then, the Gregorian calendar has been used, which tweaks for the inaccuracy of the Julian calendar by omitting the leap day at the end of three out of four centuries – just as Bede had suggested would be needed.

As the earliest indication of scientific thought in Britain, Bede's works were highly influential. *De temporium ratione* found an eager audience at home and in continental Europe, where it sparked an interest in *computus*, the calculation of the date of Easter, one of the most important considerations of the Christian Church. Even 200 years later the Church still felt a debt to Bede, a Swiss monk called Notker the Stammerer writing that 'God, the orderer of natures, who raised the Sun from the East on the fourth day of Creation, in the sixth day of the world has made Bede rise from the West as a new Sun to illuminate the whole Earth'.

Bede's works might have been lost to future scholars had not Alfred the Great, the first king of all Anglo-Saxon England, stepped in during the ninth century to ensure their survival. Although Alfred made his reputation as a masterful military tactician and courageous guerrilla warrior who ended Viking advances into southern England, he was also a learned man who earned his epithet 'the Great' as much for his educational reforms as for military achievements.

As a child, Alfred committed tracts of Anglo-Saxon poetry to memory. When he succeeded his brother to the throne in 871, Alfred taught himself to read and write, then mastered Latin. Concerned that his subjects should have access to learning in the new era of peace and stability, he went on to translate several Latin works into Anglo-Saxon, including those of Bede and Boethius, the Roman philosopher who had written extensively on ancient Greek science.

With Bede's texts preserved, the next leap in scientific investigation in Britain came shortly after the Norman invasion of Britain in 1066.

At this time, most of Europe was extremely ambivalent about science. It was an intensely theological period with a great suspicion of anything that appeared to contradict Christian teaching. Scientific and mathematical activity had shifted to the Middle East, where scholars drew on ancient Greek texts acquired following Muslim invasions of former Hellenistic cities in the seventh century. Muslim trade with Chinese and Hindu merchants, and the sharing of a common language throughout the Arab Empire, led to an Islamic Golden Age in which engineering, astronomy, medicine, mathematics and science flourished.

By placing far greater emphasis on experimental investigation than the Greeks, Muslim scientists pioneered the development of an early scientific method. In particular, Ibn al-Haytham conducted a series of experiments on optics, but the key figure was Muhammad al-Khwarizmi, a Persian astronomer and geographer regarded as the greatest mathematician of his day. Al-Khwarizmi developed algorithms and algebra (which derived their titles from his surname and from the beginning of the title of *al-jabr*, one of his publications) and adopted Hindu numerals including zero to create what we now know as Arabic numerals. When he subsequently developed the concept of the decimal point, al-Khwarizmi made long division and modern arithmetic possible.

By the early eleventh century, details of these Muslim discoveries were arriving in Western Europe via Islamic Sicily and Spain. Canute, the Viking king who reigned from 1016 to 1035, had established a tradition among English monarchs of appointing clerics from Lotharingia as bishops and masters of schools. Lotharingia was a short-lived kingdom that stretched from the North Sea coast of modern-day Holland through parts of eastern Belgium, Luxembourg and Rhineland Germany to what is now the French province of Lorraine. Contacts with Islamic Spain and Sicily that stretched back to the ninth century had earned

Lotharingian schools and monasteries a reputation among English kings as being the best in non-Islamic Europe. Canute and his successors, including Edward the Confessor and Harold II, appointed Lotharingian scholars to the bishoprics of Exeter, Hereford and Wells.

Following the Norman invasion, William the Conqueror continued the tradition of looking to Lotharingia for candidates. Among these clerics was Robert the Lotharingian, who was appointed Bishop of Hereford in 1079. Educated at Liège Cathedral school, one of the few places in northern Europe that specialised in mathematics, Robert brought his knowledge of the use of the abacus to Britain and became a pivotal figure in turning the West Country into a centre of natural philosophy in Britain. Because of his knowledge of Islamic mathematics, Robert is thought to have been appointed a Domesday commissioner by William I to survey the contents of his newly acquired kingdom.

An equally significant arrival to the West Country from Lotharingia was a noted astronomer and mathematician called Walcher. Appointed Prior of Malvern, Walcher was fluent in Arabic and knew how to use an astrolabe, a type of celestial calculator, often highly ornate and beautifully decorated, that could be used to locate and predict the positions of the Sun and Moon as well as various planets and stars.

In Italy, on 18 October 1092, Walcher watched as the Moon passed in front of the Sun to turn daylight into total darkness. A few weeks later, he heard that the same solar eclipse had been observed at his Malvern priory an hour or so earlier than in Italy. Intrigued by the implications of this observation, Walcher became fascinated by astronomy. It led him to calculate and publish a set of lunar tables that gave the time of new moons until 1111, the first mention of the use of an astrolabe in a Latin text and one of the first Western uses of Arabic astronomical data.

Walcher wrote a second text based on discussions with his teacher, Petrus Alfonsi, a scholar and translator born to Jewish parents in the north-eastern Spanish city of Huesca. Until its capture by Peter I of Aragon in 1096, Huesca had been part of Islamic Spain, and Alfonsi had been educated there in Islamic science. He translated a complete set of al-Khwarizmi's astronomical tables, the first evidence of their existence in the Latin

West and, having served as physician to King Alfonso VI of Castile, he is believed also to have been a court physician for a short period to Henry I. His arrival in Britain was a significant turning point in the flowering of Islamic science in the West Country in the 1120s. Alfonsi taught Walcher the use of degrees, minutes and seconds, which he used in a text on the times when the Moon's orbit crosses that of the Sun.

From 1125 until the end of the twelfth century, a string of English clerical scholars continued to bring Islamic science and mathematics to Britain, in most cases after they visited the East, where Muslim Europe's two greatest cities, Constantinople and Cordoba, had a wealth, splendour and vitality far exceeding anything in Christian Europe.

One of the most significant of these travellers was Adelard, one of the most colourful characters of the Middle Ages. Adelard settled in Bath in around 1135 after a journey through Europe and Asia that took in studying at Tours in France and teaching at Leon in Spain before a seven-year odyssey that he described as devoted to the 'studies of the Arabs'. Having travelled as far as Syracuse in Sicily and Antioch in the south-east corner of modern Turkey, in Tours Adelard met a well-known but anonymous wise man who explained astronomy to him. In southern Italy he spent time with an expert in medicine and nature.

The obvious differences between Islamic Europe and contemporary medieval England would have been very apparent to travellers such as Adelard. Cordoba, the capital of Moorish Spain, had wide, well-maintained and illuminated streets and squares with fountains, a low incidence of crime, a prototype police force, and over 900 public baths – more than could be found in all the countries of Christian Europe even 800 years later. Cordoba's libraries contained hundreds of thousands of books, many Moorish peasants could read and write, and the land on which they worked was highly productive. By contrast, pavements were absent from London's narrow streets until the fourteenth century, street lighting did not appear until the seventeenth century, and unsanitary conditions would play a major role in the spread of the Black Death. Outside the few towns and cities in England, vast tracts of agriculturally valuable land lay uncultivated and many peasants lived in primitive hovels, suffering malnutrition or starvation. Learning was confined to a small number of monks in monasteries drawing their knowledge from libraries that held only a few dozen books, the last refuges of culture in a period of ignorance and barbarism.

It is unsurprising, then, that Adelard despaired from abroad at his homeland, writing that 'violence ruled among the nobles, drunkenness among the prelates, corruptibility among the judges, fickleness among the patrons, and hypocrisy among the citizens'. Faced with 'this moral degeneration', Adelard added that he would seek solace in '*Arabum studia*'.

The result of Adelard's Arab studies and his seven-year journey was *Quaestiones naturales*, in which he lauded Islamic intellectualism, bemoaned the state of scientific investigation in England, and explored scientific issues of the day ranging from the pattern of tides and the structure of stars to fear of death and why fingers are of unequal length. Although some of the questions might appear frivolous, Adelard distinguished himself in the originality of his thinking and the rigour with which he applied consistent physical principles to answering them. He questioned the shape of the Earth (believing it to be round when most thought it flat) and he poured scorn on the common wisdom that the Earth remained stationary in space with the stars orbiting around it. His discussion of how far a rock would fall down a hole drilled through the centre of the Earth investigated the concept of centre of gravity. His belief that matter could not be destroyed was an early contemplation of the law of conservation of matter, and his observation that water often resists flowing out of a container when it is turned upside down involved the concepts of atmospheric pressure and vacuums. In raising all these questions, Adelard was considerably ahead of any of his contemporaries.

In a second work, *De opere astrolapsus*, Adelard turned to cosmology, the use of the abacus and the astrolabe, and spherical geometry. As a pioneer in introducing Arabic mathematics to England, he then translated Euclid's *Elements,* which later became one of the first mathematical works to be printed after the invention of the printing press and second only to the Bible in the number of editions published. He also translated many important Arabic scientific works on astrology, astronomy and philosophy.

Several other scholars followed in Adelard's footsteps, including Robert of Ketton, who studied at the Cathedral School of

Paris before travelling through the eastern Mediterranean and settling in Spain, where he translated scientific and theological texts from Arabic into Latin.

Robert of Chester also worked in Spain, where he translated Arabic books on alchemy and algebra into Latin, making a mistake that lives on today when he translated the Arabic word used to describe the ratio in a right-angled triangle of the length of the opposite side to the length of the hypotenuse, namely sine. Like the decimal system and much of modern mathematics, the name of the ratio originated in India, where the Hindus called it *jiva*. Arab mathematicians translated this as *jiba*. However, Arabic script consists of consonants with vowels punctuated underneath, so the vowels were often omitted, as when Robert of Chester came to translate the word, which he would have read as *jb*. Unaware of the Hindu origins of the word, he added in vowels that he believed to be missing, thereby yielding *jaib*, the Arabic term for a bay or an inlet, a more obvious derivation than the technical term *jiba*. Robert then translated *jaib* as sinus, the Latin word for an inlet. It stuck, and ever since then the word has been used to describe the trigonometric function, sine. Like Robert of Ketton, Robert of Chester later returned to England, arriving some time before 1147, when he calculated astronomical tables for the longitude of London.

The next significant figure in early English science was Daniel of Morley, who travelled initially to France for education. Frustrated at the paucity of the curriculum at Paris, Daniel hotfooted it to Toledo, which had gained a reputation for its rich scientific discourse. In Toledo, Daniel met Gerard of Cremona, one of the most famous translators and philosophers of his day. In 1175 Daniel returned 'with a pile of books' to England, where he found to his disappointment that English schools were no better than those he'd encountered at Paris. On the instigation of his patron, the Bishop of Norfolk, Daniel decided instead to devote his time and efforts to writing *Philosophia*, two volumes dealing with man, the creation of the world, matter, the elements, the nature of the stars and the usefulness of astrology. Daniel based his work on Adelard's *Quaestiones naturales* and several earlier Latin translations of Greek and Arabic texts. This period was the beginning of scholasticism in science – the developing of ideas and theories solely through the analysis of scholarly texts – but there were also notable examples of original ideas being developed and observations made. In 1176 Roger of Hereford wrote an ecclesiastical computus for calculating the church calendar and in 1180 Alexander Neckam laid down the first Western description of the use of a mariner's magnetic compass for navigation, which, with the printing press and gunpowder, would play a key role in triggering and shaping the Renaissance.

Intellectual discourse and philosophical investigation were given a further boost in the twelfth century by the establishment of Britain's first university at Oxford, which by the late tenth century had become a prominent town due to its position on a north–south Roman road at a point where oxen could ford the Thames – hence its name. As a frontier town between the kingdoms of Mercia and Wessex, early eleventh-century Oxford hosted a series of councils between English kings from the west and Danish invaders from the east. By the time of the Norman invasion in 1066, Oxford's strategic position had propelled it into one of the largest towns in England, boasting several thousand houses and eleven churches. Following the invasion, its location became less politically significant and Oxford's prominence declined. Trade with London in wool and cloth dwindled and more than half the properties in the town fell into 'waste'. By the mid-twelfth century, Oxford was moribund. But within its 15ft stone wall, built four years after the Norman conquest, three small monastic schools of learning, including a community of Anglican monks living in a former nunnery built in the grounds of Christ Church cathedral, had grown up. These scholars would prove to be Oxford's financial saviours.

In 1167, England was at war with France and the French king, Louis VII, expelled all foreigners from the University of Paris. Henry II encouraged the returning scholars to congregate around the monastic schools in Oxford. Formal lectures soon started and other scholars, including Dominican and Franciscan monks, were drawn to the town, bringing new prosperity to counter the declining textile trade.

By the early thirteenth century, the gathering of scholars at Oxford was formally acknowledged as comprising a university, attracting more students and beginning a long history of conflict between students and the local population that came to a head in 1209 when the townsmen hanged two students for the alleged murder of a woman, a charge that they refuted. The university dispersed, several postgraduate scholars decamping to resume their studies at Cambridge, where there was already a small school but not a recognised university.

At Oxford, the university re-formed in 1214, but clashes between town and gown continued through most of the thirteenth century. In response to riots and attacks by townsmen, the university formed primitive halls of residence to protect students, which led to the formation of the first three colleges – Merton, Balliol and University College – between 1249 and 1264. Meanwhile in Cambridge, the scholars had organised their living quarters into hostels with a Master in charge and their overall body represented by a Chancellor. As in Oxford, there was friction between rowdy students and the townfolk, who had a reputation for over-charging for accommodation. In 1231, a royal charter provided protection for students, as well as exempting the university from some taxes, allowing it to discipline its members and to monopolise teaching within the town, which was undertaken by post-graduate 'Masters' of the university rather than appointed professors from external institutions. Peterhouse, the first college, was founded in 1284.

With two universities founded, scholastic investigation of the natural and physical worlds reached its peak in Britain in the early thirteenth century when Robert Grosseteste became in effect the first chancellor of the University of Oxford. One of the leading intellectuals of his time, Grosseteste later became Bishop of Lincoln, but before then he wrote several scientific texts over an approximately fifteen-year period from about 1220. Grosseteste covered a vast array of subjects including astronomy, light and the rainbow, tides and tidal movements, the application of mathematics in science, and possibly the first scientific consideration of the origin of the universe. But what set him apart from his predecessors was a leap in understanding and philosophical reasoning that put him ahead of any of his contemporaries or any earlier medieval scientists.

The leap occurred in Grosseteste's commentary on Aristotle's *Posterior Analytics*, in which he became the first scholastic thinker in Europe fully to comprehend Aristotle's vision of the dual path of scientific reasoning. This process involved using particular observations (for instance, measurements of tide heights at various phases of the Moon) to arrive at a generalised universal law about the physical world that could then be used to make further predictions (in this instance, future tide patterns). In defining this process, which he called 'resolution and composition', and in stipulating that both paths should be verified through experimentation, Grosseteste established a key tenet of Western scientific methodology: observation leading to a hypothesis, and then confirming that hypothesis by testing it to falsification in controlled experiments.

According to his writings, Grosseteste applied this methodology only once. That was in an experiment to investigate claims that a form of bindweed called scammony caused the discharge of red bile. After excluding other known causes of the discharge, he fed someone scammony and observed the results. In so doing, he pioneered the process of experimental testing that Gilbert would later insist was the only way to gain scientific knowledge. But unlike Gilbert, who insisted that nothing could be taken for granted if it had not been tested by experimentation, Grosseteste regarded his process of resolution and composition as only one of several ways of arriving at knowledge. Nevertheless, it was a first sign that scientific thought was moving on from theoretical philosophising to practical experimentation.

Grossseteste's second and equally important contribution was his belief in a concept of the subordination of sciences, in which, for instance, optics would be subordinate to geometry because it relies on geometry. At the top of Grosseteste's hierarchy of sciences was mathematics, which led to the suggestion that no scientific observation could be regarded as valid if it could not be described in mathematical terms.

Grosseteste's writings were enormously influential and would be drawn upon for much of the next century. For many scholars and intellectuals, simply citing a Grosseteste quotation would often be enough to defend an opinion or to assert a supposed fact, even when Grosseteste's teachings were not fully understood. One case in point was a Franciscan monk called Bartholomew the Englishman, or Bartholomaeus Anglicus as he was known by his order, who did not let his lack of understanding stop him from citing Grosseteste's speculations on the nature light when compiling *De proprietatibus rerum* (*On the Properties of Things*), the most widely read and quoted encyclopaedia of the Middle Ages.

A student of natural sciences and theology under Grosseteste at Oxford, Bartholomew covered all the sciences known at the time he completed his nineteen-volume opus in about 1245. Various volumes summed up the contemporary thinking on medicine, astronomy, chronology, zoology, botany, geography and mineralogy, among others. Its encyclopaedic range and clear presentation made his work an immediate and widespread bestseller, bringing celebrity to Bartholomew, who became known as the 'Master of Properties' in the wake of its success. Translated into several languages, including English, it was popular for several centuries and in 1472 became one of the first encyclopaedias to be printed, appearing in Cologne only eighteen years after Johannes Gutenberg invented the movable-type printing press.

Grosseteste's extensive writing and teaching on natural philosophy at Oxford prompted another of his students, Roger Bacon, also to attempt to compile an encyclopaedia that encompassed all knowledge of the natural world. But unlike his predecessor Bartholomew, Bacon's attempt to provide a rational, empirical explanation ended not in his celebrity but in his imprisonment.

Born in Somerset in the early thirteenth century, Bacon was passionately interested in the natural world and how things worked, and was in many ways an English equivalent of Leonardo da Vinci. Remarkably driven compared with his contemporaries, heavily influenced by spirituality (he became a Franciscan monk in his forties) and outspoken to a degree that risked his freedom and well-being, Bacon became known soon after his death as 'Doctor Mirabilis' on account of his alleged interest in alchemy and magic. However, like many similar accusations of sorcery, this suspicion was more the result of a reactionary fear that his discoveries undermined the control of the Church than any belief by Bacon in the occult.

Like da Vinci, Bacon envisaged future technologies centuries before they were possible. In the thirteenth century, some 200 years before da Vinci had similar ideas, Bacon mentioned the possibility of carriages propelled without the use of animals. He also envisaged self-driven boats, an 'instrument small in size, which can raise and lower things of almost infinite weight', microscopes, telescopes and possibly a pinhole camera. Bacon is the first person known to have thought of buoyant flight, suggesting that it could be achieved by filling a thin-walled metal sphere with rarefied air or liquid fire, thereby anticipating hot-air balloons and gas-filled airships. But unlike da Vinci, whose reputation as a Renaissance scientist or technological visionary relied solely on the theoretical sketches and jottings found in his notebooks, Bacon was a doer and an experimentalist.

Bacon's experimentalism began with gunpowder, for which he is often credited as the inventor. However, it is more likely that gunpowder was invented in China and that Bacon learnt of it from Arabic writings, which he read well. Certainly, he experimented with various proportions of black powder, as it was called because it resembled soot, using the ingredients saltpetre, sulphur and coal. He was so impressed with his results that in 1242 he concealed them in a coded Latin anagram that was not cracked until the twentieth century. Unfortunately for Bacon, he did not manage to obscure fully the nature of his mixture or its explosive properties, which rapidly became general knowledge, having a dramatic and decisive impact on the course of civilisation and on his own liberty. In these early years of the great universities, any learning or experimentation that produced insights or discoveries that threatened the Bible's account of history could have catastrophic consequences. For any intellectual who, like Bacon, was studying in Paris at that time, the threat of mob violence and repressive religious censorship was always present. To those mobs and censors, Bacon's experimentation with powders to produce an explosive mixture seemed like diabolical dabbling and he was confined to a monastery, ostensibly for dabbling in witchcraft.

After a decade in exile, in about 1256 news emerged that Bacon had unexpectedly become a friar in the Franciscan Order, a surprising vocation as he appeared to hold little in common with the ascetic practices stipulated by St Francis of Assisi. As a Franciscan monk, Bacon could no longer hold a teaching post at a university and his activities were further restricted in 1260 when a Franciscan statute was introduced that prohibited friars from publishing books or pamphlets without specific approval.

Nevertheless, Bacon continued his experimentation, his reading of Arabic texts prompting him to investigate optics and the refraction of light through lenses. Drawing heavily on translations of treatises by the great Arabic scientist Ibn al-Haytham, who had flourished in Egypt 250 years earlier, Bacon became the first Western thinker to understand the finer points of al-Haytham's leading ideas, including his assertion that vision took place by coloured light entering the eye rather than visual rays being projected out of it. Al-Haytham also described the construction and operation of parabolic burning mirrors, which played a part in directing Bacon's experimentation with lenses. As a result of these experiments, Bacon discovered the magnifying properties of a convex lens, as well as the many distortions that could be achieved with transparent shapes of glass and other materials. In doing so, he invented the magnifying glass and suggested the use of spectacles for far-sightedness, but, as with his work using gunpowder, Bacon's feats with shaped glass were regarded with misgiving by adherents of the Church. If they were possible at all, it was thought, they could not be natural and godly, and must therefore be illicit. Bacon again fell under suspicion of sorcery and necromancy.

Bacon would most probably never have written of his discoveries – and might have been imprisoned again, or worse, for his optical experiments – had he not several years earlier met Cardinal Guy le Gros de Foulques, a former lawyer and military man who was secretary to King Louis IX of France. By 1261 Guy had become a cardinal. Interested in Bacon's schemes, Guy encouraged him to investigate further. Then in 1265 Guy was elected Pope Clement IV and promptly issued a mandate ordering Bacon to write to him concerning the place of natural philosophy within theology.

The papal mandate brought about Bacon's most famous work, the *Opus majus*, an encyclopaedia of knowledge, on everything from alchemy to optics. Written in secret to avoid the opposition of Bacon's superiors, the synopsis alone took Bacon a year and ran to 800,000 words. Divided into seven parts, it attempted to summarise (albeit at some length) the entire gamut of natural knowledge.

Essentially a discussion of how the philosophy of Aristotle and experimental science could be incorporated into contemporary theology, the first three parts of the *Opus majus* served as an introduction to the ideas that would be discussed in detail in the subsequent parts: the causes of human ignorance, the relation of the other sciences to theology, and grammar and the power of languages. The subsequent parts then attempted to explain mathematics, astronomy, astrology, optics, the concept of experimental science and moral philosophy.

As soon as it was finished, Bacon handed the *Opus majus* to a messenger with instructions to carry it immediately to Rome. But shortly after his treatise had been dispatched, Bacon started to worry. Concerned that his work might be lost in transit or that it would be too long for a busy pope to read, and feeling he'd left some things out that he now thought should have been included, he barely paused for breath before starting on another work, the *Opus minus*, in which he intended both to summarise and to supplement his previous, larger work.

Bacon's *Opus minus* was soon finished and, like his *Opus majus*, dispatched by messenger to Rome, possibly with an additional treatise on logic called *De multiplicatione specierum*, other works on alchemy and astrology, a text on burning mirrors called *De speculis comburentibus*, and an optical lens. But even then, Bacon wasn't satisfied that he'd done his subject justice and consequently embarked on his *Opus tertium*, which was considerably longer than either of its predecessors.

It is likely that the *Opus tertium* was neither properly finished nor sent to the pope, as Clement IV died in 1268, fewer than four years after his coronation. The pope left no record of his response to Bacon's ideas, if he read the treatises at all. Bacon promptly lost his papal protection, putting him at the mercy of the many enemies he'd attracted as a bold, outspoken and frequently highly critical man. To make matters worse for Bacon, he'd also written a significant work on the calendar, called *Computus*, in which he proved errors in the Julian calendar and sought its reform. Bacon's outspokenness could not have been worse timed. At that time the Church was ruthlessly pursuing anyone who questioned the calendar, fearing that this might undermine its infallibility.

Nevertheless, Bacon abandoned neither his writing nor his experimentation following the death of Clement IV, although he did express disillusion and fury at what he regarded as the increasing corruption of knowledge. In 1277, Jerome of Ascoli, the

Minister-General of the Franciscan Order, declared himself against Bacon's books and forbade the reading of them. In the same year, on the prompting of Clement IV's successor, Pope John XXI, the bishops of Paris conducted investigations into the teaching of Aristotelian science at the University of Paris. The outcome was the Condemnations of 1277, which banned the teaching of certain philosophical doctrines, including deterministic astrology. Bacon's interest in alchemy, optics and calendar reform fell foul of the ban and led to claims of sorcery. In 1278 he was placed under house arrest and accused of heresy, eventually moving from his Paris quarters to the Franciscan House at Oxford.

Bacon's house arrest was lifted just months before he died in Oxford in 1292. Although his *Computus* was cited in pleas to Elizabeth I for calendar reform in the sixteenth century, Bacon's *Opus majus* was not published until 1733, by which time Pope Gregory XIII had appointed the commission that gave us the calendar we have today.

Credited with being the first person to put forward the idea of experimental investigation – that is, investigating and recording the effect of successive small changes under experimental conditions in order to gain a mechanical understanding of the secrets of nature – Bacon has been regarded by some historians as the first scientist. But there is some ambiguity, as Bacon termed his method a 'science of experience', by which he could just as well have meant intellectual experience as experimental observation. In his favour, it appears he was one of the first people to use experimental methods in alchemy, the root of modern chemistry. And as mentioned above, he applied geometry to the science of lenses and made early experiments in gunpowder. He also achieved geographical breakthroughs later used by Columbus, and stressed the importance of mathematics to science, a significance that would not be fully recognised for 400 years.

Like Gilbert, Bacon believed that observation and exact measurement were the surest foundations for truths about the natural world. But unlike Gilbert, he did not advocate abandoning all beliefs that had not been tested by experimentation. Many of his ideas were born of scholastic investigation and, like Grosseteste, he retained a belief in astrology and psychic influences, such as that words have mystical powers when they are spoken in certain psychological mindsets. These caveats notwithstanding, at the very least Bacon played a significant part in promoting a philosophy of experimental investigation, and his many treatises, although unpublished at the time, were a sign that a new philosophy of experimental investigation was now starting to flower.

It was no coincidence that Bacon chose Oxford as the site of his house arrest and his continuing scientific investigation. In the time he'd been in Paris, Oxford's tradition of scholastic examination of natural philosophy texts had continued apace and within a few years of Bacon's death, in about 1294, Merton College had become the focus of a group of dons that in the early fourteenth century rivalled the scholastics at the University of Paris.

Merton, the first fully governing college in the university, had only recently been founded by Walter de Merton, a former Chancellor of England and Bishop of Rochester. For several decades of its early life, the college was largely a building site. Several buildings that now surround Mob Quadrangle, the oldest quadrangle in the university, were built in the late thirteenth and early fourteenth centuries, including the north and east ranges and the Sacristy of Merton Chapel, construction of which began in about 1290. However, of Merton's original buildings only the chapel now remains largely unchanged in its ancient state.

With undergraduates not admitted until the 1380s, the college was a comparatively intimate and close-knit community of twenty fellows, the most notable of which were known as the Merton or Oxford Calculators. This band of scholars applied mathematical or logico-mathematical methods to questions of natural philosophy (mainly physics, mathematics and astronomy) and developed techniques that remained influential until the time of Newton. Their main contribution was to separate the classical mechanics of moving bodies into two separate disciplines. Kinematics was concerned with the motion of a body – for instance velocity, acceleration and rotational movement – while ignoring the causes of that motion. The other branch, called dynamics, looked at the causes of motion of a body, such as inertia and mass.

Unlike their predecessors such as Roger Bacon and Roger Grosseteste, the Merton Calculators' work was entirely theoretical and contained neither empirical nor experimental elements. Instead it relied on using supposition theory and logical exposition to solve mathematical puzzles, a process that would nowadays be called mathematical analysis. Through this the Calculators achieved some notable breakthroughs that influenced the development of early modern science.

Among them were the logicians Walter Burley and John Dumbleton, a future Archbishop of Canterbury called Thomas Bradwardine, and Richard Swineshead, who was regarded as the Calculators' leading light and one of the most outstanding intellects of his time. Known simply as 'The Calculator', Swineshead examined how quantitative mathematics might be applied to concepts that were considered at that time to be non-mathematical and qualitative, such as colour and heat.

However, probably the greatest single achievement was by Bradwardine, who overturned Aristotle's suggestion that velocity is proportional to force and inversely proportional to resistance. Looking at changes in velocity (in other words, acceleration), Bradwardine found an exponential relationship. This discovery, which meant a doubling of velocity would require the ratio of force to resistance to be squared and a tripling of velocity would require that ratio to be cubed, was one of the first significant challenges to the great Aristotle's intellectual hegemony.

Another calculator, William Heytesbury, was a doctor of theology who at the end of his career became Chancellor of the University of Oxford. With some of the other Merton Calculators, Heytesbury had developed theories concerning the logic of continua and infinite divisibility (the philosophical question of whether something can be infinitely split into smaller parts), but his major accomplishment was to formulate a mean speed theorem long before Galileo, who is generally credited with exactly the same theorem in his law of falling bodies. The mean speed theorem states that a body travelling at constant velocity will cover the same distance in the same time as an accelerated body if its velocity is half the difference between the starting and finishing speed of the accelerated body. Thus someone already moving at 10 m.p.h. would be caught up by a second person accelerating smoothly from a standstill as soon as the second person reached 20 m.p.h.

Although the formulation of this theory was an end in itself to the Merton Calculators, it would eventually lead to the development of fundamental calculus theory by Newton, enabling him to unearth deep and fundamental truths about the world, about how things moved and about what seemed to be an almost magical force – gravity – that influenced all objects.

At the same time as the Calculators were in full flow at Merton, a Franciscan monk who had studied at the University of Oxford from 1309 to 1321 was undertaking a scholastic analysis of texts in logic and natural philosophy that would have profound implications on science and philosophy that still apply in the present day.

Waiting for an opportunity to undertake his doctorate at Paris or Oxford, William of Ockham was lecturing on Aristotle's logic at a Franciscan school and writing analyses of several key texts, including on Aristotle's *Physics*. Named after the small village in Surrey where he was born around 1280, Ockham was called 'doctor invincibilis' by his fellow Franciscans in recognition of his intellectual powers. This invincible teacher proposed many new ideas in his writings, among them an intriguing philosophical principle that has become known as Ockham's razor. Also known as the principle of parsimony, it states that 'plurality ought not to be posited without necessity'. In essence, this maxim of 'simpler is better' advises economy and the avoidance of complication when seeking explanations of the workings of the world. In other words, one should always opt for an explanation in terms of the fewest possible number of causes, factors or variables.

Ockham's razor had profound implications for the development of the scientific method and is cited to this day by scientists, including Stephen Hawking, when developing theories and testing their arguments, particularly in the fields of physics and astronomy.

Less than 200 years later, the principles embedded in Ockham's razor would be demonstrated in the success of the Copernican model of the solar system over the Ptolemaic model. The Roman philosopher Ptolemy put the Earth at the centre of the universe with the Sun, planets and other stars orbiting around it. However, Roman astronomers had noticed that Venus appeared to orbit the Earth faster than Mercury. To explain the apparent backwards motion of Mercury relative to Venus, adherents of Ptolemy suggested that Mercury's orbit spun in small circles which moved along a larger circular orbit that stretched around the Earth, a pattern called an epicycle. This convoluted explanation, which also involved the planets being embedded in crystalline celestial spheres, just about worked, but it seemed extremely complex when compared to Copernicus' and Kepler's elegant explanation that put the Sun at the centre of the solar system and the planets on elliptical orbits around it. A classic example of a phenomenon that conformed to Ockham's razor, the simplicity of the Copernican model eventually won the day and was later proven by experimental investigation.

In the nineteenth century, Ockham's razor again proved its worth. At that time, physicists believed that light, like sound, required a medium of transmission, so they came up with a theory that a 'universal ether' performed the role. Even though many experiments failed to find evidence of an ether, scientists continued to assume it existed. That is, until Einstein's theory of special relativity provided a much simpler and considerably more elegant explanation.

Ockham's razor was also applied to branches of biology, although it is less applicable to complex systems that are shaped by, say, environmental or genetic factors. However, medics often apply a similar rule, called the principle of diagnostic parsimony, which advocates applying the fewest possible causes to explain many different symptoms, although they are also made aware in their training that several simple causes will often be a more likely explanation than a single complicated or rare disease.

Like many medieval natural philosophers, Ockham eventually fell foul of religious authorities. Shortly after devising his principle of parsimony, he was summoned before the papal court at Avignon to have his writings examined for heresy. Ockham's treatise on the character of knowledge – the difference between the name of a thing and the thing itself – had caught the eye of Pope John XXII, who in 1324 appointed John Lutterell, a former Chancellor of Oxford, to the task of examining Ockham's entire works. Lutterell's initial examination found fifty-six statements considered to be heretical. A second commission of theologians was appointed to inspect Ockham's work in greater detail, but before the commission could report, Ockham and several fellow fundamentalist Franciscans counter-attacked Pope John XXII, accusing him of heresy over the Church's views of the poverty of Christ and the apostles. Fearing imprisonment and possible execution, the Franciscans fled Avignon, eventually arriving in Pisa, where they took refuge in the company of Emperor Ludwig of Bavaria, the leading political opponent of the pope. Ockham was excommunicated and spent the last years of his life at a Franciscan convent in Munich, obsessing over the implications for the Church of a heretical pope.

Ockham never returned to England, which by the time he died in about 1350 was in the grip of the Black Death. The rich tradition of Oxford scholastic philosophy declined at about the same time, mainly as the result of the devastating effect of the plague on primary education, which left the next generations of students insufficiently prepared to improve upon the work of their predecessors. As the Black Death tightened its hold on Europe over the next three centuries, English universities suffered worse than their continental counterparts as they attracted fewer foreign students to offset the sudden decline in student numbers. By 1375, when the Black Death was at its most virulent, average life expectancy had dropped to just over seventeen years, bringing a sudden closure to a long period of massive scientific change from which British science did not reawaken until the Renaissance moved to northern Europe in the late sixteenth century and the invention of printing democratised learning and allowed a faster propagation of new ideas.

Chapter 2

THE FIRST SCIENTISTS

After the Black Death swept through Europe, William Gilbert's experimental investigations of magnets and electrical forces at the beginning of the sixteenth century were the first significant sign for 250 years of a renewed interest in science in Britain. However, Gilbert's ideas did not emerge from an intellectual vacuum. As well as drawing on the ideas of earlier generations of British thinkers, Gilbert supported the Copernican view of the universe and was therefore highly influenced by the Renaissance.

The initial mid-fourteenth century period of the Italian Renaissance had resulted in a marked lull, and possibly a backwards step, in scientific progress. The rise of Renaissance humanism viewed nature as a spiritual creation at odds with the laws of physics or mathematics. And although the humanists' emphasis on the pursuit of a good and active life (over a contemplative life) fostered a belief that a knowledge of nature could be useful, interest in philosophical investigation had waned as intuition and emotion replaced logic and deduction. Long after several generations of scholastic natural philosophers and early experimentalists such as Roger Bacon had questioned classical world views, Renaissance scholars returned to the texts of Aristotle and Ptolemy, which they regarded with a new reverence. As a result, there were no new developments in physics or astronomy.

However, by the mid-fifteenth century, when the Renaissance moved to Northern Europe, several events led to a revival of scientific enquiry that would eventually make itself felt in England by the early sixteenth century. The first of these was the fall of Constantinople to the Turks in 1453, which sent many Byzantine scholars fleeing for refuge to the West, particularly to Italy but also on to Germany. These scholars brought with them ancient texts that were seized upon by the Italian humanist movement, which yearned to return to the height of civilisation last seen in Western Europe at the end of the Roman Empire. In the Byzantine texts, the Italian humanists found alternatives to the all-pervasiveness of Aristotelianism. Plato and other philosophers stimulated a renewed interest in logic and deduction, as well as an appreciation of the role of mathematics, which Aristotle always downplayed as a means of gaining knowledge. This reappraisal of mathematics was the first of three crucial turning points that led to the scientific revolution of the mid-sixteenth century.

The second factor was the development of new technologies that, unlike science, could initially be developed by trial and error in response to a need. These technologies not only stimulated scientific investigation and experimentation (in some cases in order to improve the technologies) but made new types of science possible, and so a beneficial circle between new technologies and new science was established. The most significant of these new technologies appeared a year after the fall of Constantinople, when in 1454 at Mainz in Germany Johannes Gutenberg printed the first copies of his Vulgate Bible on his movable-type printing press. This symbolic first step in the spread of the printed word democratised learning and led to a much faster propagation of new ideas. Equally significant was the development of long-distance sailing ships, which brought back knowledge from abroad, and after 1492, when Christopher Columbus discovered the Americas, stimulated a general European awareness of a new world that led to greater interest in geography, navigation, the compass, astronomy and exploration.

The third factor was the dramatic depopulation of Europe by the Black Death, which encouraged the development of new technologies to deal with the loss of manpower and changed social, political and cultural attitudes. The senseless, indiscriminate ravages of the plague led the survivors to question everything; not just the relevance of science and philosophy but also the role of religion and the Church, which underwent the great upheavals of the Protestant Reformation of Martin Luther and John Calvin, and subsequently the Catholic Counter-Reformation. In this period of massive theological and cultural change, a suitable environment for progress evolved in which it was just as legitimate and equally valid to question scientific teachings as religious doctrine.

If any single event in this period of upheaval could possibly be cited as catalysing the start of the scientific revolution, then arguably it was the Catholic Counter-Reformation.

Seeking to counteract the Protestant Reformation, the Catholic Church convened one of the most important ecumenical councils in its history at Trento in Italy from 1545 until 1563. In addition to announcing condemnations of what it regarded as Protestant heresies, the Trent Council issued numerous reform decrees, many of which were intended to make the Catholic Church more attractive to its potential followers. One of its last decrees was a decision to investigate reform of the Julian calendar, which by 1563 had put Easter ten days adrift from the seasons.

Among those asked to work on the problem of calendar reform was Nicolaus Copernicus, a canon at Frombork in northern Polish Prussia. Twenty years earlier, Copernicus had published *De revolutionibus orbium coelestium (On the Revolutions of the Heavenly Spheres)*, the treatise that would eventually be recognised as a masterpiece but which had so far attracted little attention. In its introduction, Copernicus explained that the need for calendar reform was one of the reasons he had decided to investigate a proper measurement of the length of the year. That investigation led him to develop his heliocentric model that directly threatened the Ptolemaic view of the solar system, the only explanation of the solar system accepted by the Church. In *De revolutionibus*, Copernicus directly contradicted literal readings of the Bible and unleashed an idea that would eventually bring countless scientists, from Galileo to Newton, into conflict with the Church in the seventeenth century. In the meantime, however, the Church had failed to notice Copernicus' potential heresy and in 1582 Pope Gregory XIII introduced a new calendar. Initially the change was made only in four Catholic states, as many Protestants resented what they saw as a Catholic invention designed to return them to the Catholic fold.

Copernicus had set the ball rolling. Unwittingly, the Catholic Church had allowed its quest for calendar reform to question their orthodox doctrine. Ptolemy's works of astronomy had been found to be inconsistent with everyday observations. Realising that the classical philosophers were after all only human, Renaissance scholars now had the confidence to move on from antiquated worldviews. Now curious minds felt empowered to challenge anything, particularly as another classical scholar had already been toppled: Galen was a Greek physician whose writings on anatomy, with its four humours of the body, were revered by the Renaissance humanists. His 'assassin' was Andreas Vesalius, born in Brussels into a medical family.

Vesalius studied medicine in Paris and at the University of Louvain with a fervour that bordered on the criminal when he stole a body from under a gallows outside Louvain and dragged it home for study. After graduation, he moved on to the University of Padua in Italy, where he was aided in his obsession with dissecting human cadavers by a judge who would delay the times of executions of criminals to suit Vesalius' study schedule. With so much hands-on experience, Vesalius discovered around 200 errors in Galen's accounts of anatomy, including the crucial fact that the wall separating the left and right heart ventricles was not perforated, thereby immediately destroying Galen's theories on human physiology. Realising that Galen had no real experience of human dissection, Vesalius decided to write a book about his experiences.

De humani corporis fabrica (*On the Fabric of the Human Body*) was published in 1543, overturning centuries of established thinking on anatomy in much the same way as Copernicus' book, published the same year, did for astronomy. As much a manual on how to dissect as a textbook of anatomy, it was also quite beautiful. With illustrations by John Stephen of Kalkar, a leading pupil of Titian, it presented its readers with an insight into a world they had previously been unaware existed. Like Gilbert's *De magnete*, published nearly sixty years later, it also laid down a manifesto for scientific experimentalism. Classical teachings handed down through the generations could no longer be relied upon, Vesalius said, unless those teachings could be verified through personal visual experience. In other words: trust nothing if you haven't examined it yourself. Vesalius' ethos would have a profound effect on Gilbert in the years of his medical practice. It prompted Gilbert to reject the medical establishment's adherence to the philosophy of personal experience and led to his advocating of investigation and testing through experimentation. It also prompted another Englishman, William Harvey, to build on Vesalius' methodology to make discoveries every bit as revolutionary as Gilbert's achievements.

* * *

Born in 1578 to a prosperous Kent family, Harvey studied medicine at Cambridge and at the University of Padua, where he was taught by Hieronymus Fabricius, who held the chair of professor of anatomy, the same position that Vesalius had occupied. In fact, Fabricius provided the final link in a direct line between Vesalius and Harvey, as he was taught by and preceded as professor of anatomy by Gabrielle Fallopio, who first identified the tubes that bear his name and who was taught by Vesalius.

In 1579, Fabricius discovered the valves in veins, but did not explain their purpose. When Harvey arrived at Padua in the late 1590s, he became intrigued by Fabricius' discovery, eventually discovering the answer by pioneering methods that owed as much to common sense and mathematics as experimentation. In doing so, he played a key role in establishing scientific practices for all scientists.

Harvey knew he faced formidable opposition if he was going to overturn centuries of thinking on the role of blood vessels and the heart. Natural philosophers had recently challenged Galen's views, not least Michael Servetus, a Spanish physician. In 1553, Servetus had published his deduction that the heart circulated blood first to the lungs before pumping it to the rest of the body, a view that, like Vesalius' discoveries, did not fit with Galen's explanation. Unfortunately for Servetus, his book also contained theological views that were regarded by Calvin as heretical. Long before Harvey arrived at Padua, Servetus' book had been destroyed and its author had been burned at the stake.

Provided Harvey did not also venture into theological discussions, he was unlikely to experience the same fate as Servetus, but he knew he would face considerable resistance to a change in doctrine from a highly conservative medical establishment. In spite of their discrediting by Vesalius, many of Galen's ideas had survived, not least his explanation for the movement of blood. According to Galen, the liver converted food into thick, darkly coloured blood, which flowed via veins to the heart, where it somehow passed through the septum (which Vesalius had shown was not perforated). Heart spasms pumped this now lighter-coloured and thinner arterial blood to all parts of the body. Meanwhile, a second supply of arterial blood distributed some kind of 'vital spirit' from the lungs around the body. Both types of blood were consumed in the process and required constant replenishment. Even Fabricius still subscribed to this explanation, believing the valves he'd identified slowed the flow of blood from the heart so that it could be more easily absorbed by the body.

Harvey took his time. He knew he'd have to prepare a watertight case if he wanted to challenge Galen's account of the workings of the human body. Having returned to England in 1602, he married the daughter of one of Queen Elizabeth's physicians in 1604, a union that undoubtedly helped his swift ascent through the ranks of the medical profession to become in 1609 a professor at St Bartholomew's Hospital in London, Physician Extraordinary to James I in 1618 and then Royal Physician to the monarch.

In his spare time, Harvey conducted research, performing more than 100 vivisections on the bodies of at least eighty species of mammals, their still-beating hearts and blood vessels providing insights that would revolutionise our understanding of anatomy. With the support of James I and Charles I, who succeeded his father to the throne in 1625 and allowed Harvey to dissect deer from the royal parks, Harvey methodically built up his case. The king aided Harvey again in his research when he introduced him to Viscount Montgomery, who as a boy had fallen from a horse, leaving a gap in his ribs that was covered by a metal plate. 'I immediately saw a vast hole,' wrote Harvey, describing how Montgomery removed the plate to let him feel and see his heart beating beneath scar tissue.

Operating at the frontier of scientific knowledge required Harvey to make leaps of imagination and deduction with which he often struggled. 'When I first gave my mind to vivisections,' he wrote, 'I found the task so truly arduous, so full of difficulties, that I was almost tempted to think that the motion of the heart was only to be comprehended by God.'

Harvey pursued his research with an unusual zeal, always carrying two notebooks – one for theories, one for facts – and a dagger in case he came across an unexpected opportunity for a dissection. He laid out his thoughts and ideas for experiments in the theory book. He recorded his results in his fact book with meticulous descriptions and detailed drawings. Then, establishing a practice that changed the path of science by making him the first practitioner of the scientific method, he insisted on repeating every experiment. If it didn't elicit the same results, it would be removed.

Eventually his dedication in pursuing his vocation 'at length, by using greater and daily diligence and investigation, making frequent inspection of many and various animals, and collating numerous observations' paid off. At last, Harvey could write: 'I thought that I had attained to the truth.'

Although he had started to lecture on his discoveries in 1616, Harvey waited until he was 50 years old, in 1628, before publishing his conclusions in *Exercitatio anatomica de motu cordis et sanguinis in animalibus* (*An Anatomical Exercise on the Motion of the Heart and Blood in Animals*). The small book had only seventy-two pages, was beset with typographical errors and printed cheaply in Germany on thin, flimsy paper, but in it he gave full details of the experiments he had conducted. Now he had the evidence to formulate his conclusions. Writing in the introduction, Harvey explained his process of discovery by experimentation. 'I profess to learn and teach anatomy not from books but from dissections,' he wrote, 'not from the tenets for philosophers, but from the fabric of nature.'

Harvey's first revelation was that blood flowed away from the heart in arteries and towards the heart in veins. He'd discovered this by tying a tourniquet around his own arm and those of several volunteers. With the tourniquet pulled tight, he stopped blood flow in his arteries, which bulged with blood on the side of the ligature closest to the heart, thereby indicating that arteries carried blood away from the heart. Harvey then relaxed the tourniquet, so that it constricted only the flow in his veins, which were less deep in his arm. This time only the veins bulged and only on the side of the tourniquet furthest from the heart, indicating that the veins carried blood back to the heart. This experimental evidence showed that blood did not oscillate in blood vessels in the way that Galen described. Instead, said Harvey, the 'heart sends blood through the body in a loop'.

From his many dissections, Harvey had discovered that the heart also had valves, just like the veins. In this case, the valves allowed blood to flow only from the two upper auricle chambers of the heart to the two lower ventricle chambers. Harvey explained how this allowed the blood to flow only in one direction and through two circulatory loops, first the pulmonary loop from the right ventricle via the lungs to the left auricle, then the systemic loop from the left ventricle via the rest of the body back to the right auricle.

To support his theory of circulation and a closed blood supply, Harvey cited his examination of artery walls, which he'd found to be thicker close to the heart. This would be expected if arteries needed to withstand the higher pressure generated by the heart's pumping action. He also cited anecdotal evidence by doctors on the speed at which poisons circulated through the body.

But his most convincing evidence came from his dissections of hundreds of mammal hearts and the first example of a quantitative explanation in biology. Harvey had weighed human and animal hearts, measured their pumping capacities and counted the creatures' heartbeats per minute. He'd also drained animals of their blood, measured its volume and compared it with his heart measurements to determine how much blood was in total circulation and how much blood left the heart with each beat.

Intentionally underestimating volumes so that he could not be accused of exaggeration, Harvey calculated that each beat of a human heart pumped out about 60 cubic centimetres (3.7 cubic inches) of blood. Even at a relatively low pulse rate, this would amount to 260 litres (57 gallons) an hour, about three times the weight of an average man and many times more than could be feasibly consumed in Galen's model of the blood.

Galen's orthodoxy had again been shattered, but although Harvey's book later became a classic in science literature, at the time of its publication its author was widely attacked. Few readers or members of the medical profession were willing to accept the biggest change in medical theory since the discovery of the nervous system by Alexandrian physicians nearly 2,000 years earlier. The main objection was that Harvey's account of circulation had no medical purpose – neither oxygen nor its respiratory role had been discovered at that time. Some objectors dismissed

Harvey as a mere accountant whose adding up of blood quantities was no match for philosophical reasoning. Other opponents dismissed his explanation of circulation simply because it did not fit with the Galenic theory of humours of the body.

Harvey persevered with his research and lecturing – although he did his reputation as an experimentalist no favours when he responded to a demand by King Charles I for a medical inquiry into some 'witches' from Burnley. Abandoning his rational, experimentalist approach to science, he caved in to the superstitions of the day, declaring the women not to be witches because none of them sported more than the conventional two nipples.

In 1635, when asked by the king to perform an autopsy on Old Tom Parr, a national celebrity who had supposedly lived for 152 years, Harvey again abstained from applying his usually rigorous scientific methodology. Parr alleged he was born in 1483, joined the army around 1500 and married at the age of 80, when he supposedly fathered two children, both of whom apparently died in infancy. He then claimed to have had an affair at the age of 100 and, after his wife had died, apparently married for a second time at the age of 122. Painted by Rubens and Van Dyke after news of his purported age spread, his longevity would have meant he had lived through the reigns of ten kings, but this came to a sudden end when his landlord, Thomas Howard, Earl of Arundel, brought him to London in 1635 to amuse the latest monarch. When Old Parr fell dead within a year of his arrival in the capital, Harvey dissected his body and, instead of citing his obvious old age or any other medical condition as the cause of his death, Harvey declared it to be the result of his removal from his usual way of life to the hurly-burly of seventeenth-century London.

However, these episodes were small diversions from Harvey's experimentalism. Not content to rest on his laurels, he continued his research, moving on to the processes of reproduction in about 1630. Devoting another two decades of his life to researching the subject, the only interruption to his work came at the outbreak of the Civil War in 1642 when his research notes were destroyed during the ransacking of his lodgings in London riots. Although Harvey later described the loss of his notes on insect reproduction as the 'greatest crucifying', his overall response was typically phlegmatic: he vowed to re-create his experiments to produce a new set of notes. In the meantime, Harvey left the riots of London behind to accompany Charles I to the Battle of Edgehill, the first pitched confrontation of the war, where he was assigned to protecting the king's children. As the battle proceeded, Harvey hid the Prince of Wales and the Duke of York in a hedge, evading the Parliamentary forces by reading to the princes in a whisper to keep them silent. Then, when the battle was over, Harvey treated the wounded regardless of their allegiance.

In 1643, Harvey ended his association with St Bartholomew's Hospital and headed with the king to Oxford to become Warden of Merton College, retiring from public life when Oxford surrendered to Cromwell in 1646. He continued investigating fertilisation, applying the same techniques he pioneered in his study of circulation, in this instance examining the fertilised eggs of hundreds of mammals. Nine years later, in 1651, when Charles I had been defeated and beheaded, and the Commonwealth of England was ruled by the Rump Parliament, Harvey published his second major work, *Exercitationes de generatione animalium* (*Essays on the Generation of Animals*). Concerning himself primarily with the genesis of mammals, Harvey put forward the entirely new theory that mammals reproduced through the joining of an egg and sperm. From his observations of chick and other embryos, Harvey asserted that no other theory made sense. Although he had little experimental proof beyond his observations, Harvey's theory was so compelling and so well thought out that it became the basis for modern embryology, the world assuming he was right 200 years before a mammalian sperm fusing with a mammalian egg was finally observed.

In spite of his numerous groundbreaking discoveries, his belief in experimentalism and his establishment of a scientific method, Harvey did not fully deserve the title of a modern scientist. This was because he remained wedded to several concepts that were more mystical than scientific, such as a belief that the body was kept alive by spirits and vital forces. He died of a stroke in 1657 at the age of 79, but his legacy lived on in various endowments, including for a library at Merton, and his teaching of a generation of students who embraced his findings and methodology.

* * *

In gradual but sure-footed steps, British science was shaking off the ruminations of natural philosophy and replacing it with empiricism and mathematics. While Gilbert and Harvey had been pioneering experimentalism and the scientific method in England, north of the border a highly eccentric Scottish laird with a penchant for astronomy and numbers had made one of the single greatest advances in the history of mathematics.

John Napier, eighth Laird of Merchiston, was born just outside the city boundary of Edinburgh at a time of great upheaval in Scotland. By the time he started his formal education, aged 13 in 1563, Scotland was under the rule of Mary, Queen of Scots, a devout Catholic deeply resented by Protestants such as Napier, who declared his primary aim in life to be the advancement of the Calvinist cause against the 'blindnesse of papists'.

After a brief period abroad during which he gained an education in higher mathematics and classical literature, Napier returned to Scotland in 1571 and soon took over the large family estates, building a castle in which he took up residence with his recently wedded wife in 1574. For nearly twenty years, Napier busied himself with running his estates in a way that – in contrast to his narrow religious views – was scientifically enlightened, always looking for new technologies and techniques that he could apply to the land. Napier's ingenuity soon gained him a reputation around Edinburgh as 'Marvellous Merchiston'. He experimented with salt fertilisers and manures. He invented a pump that removed water from flooded coal pits, and he designed devices to better survey land. He also devised machines to defend the country from any invasion by the Catholic Spanish king, Philip II, including a precursor of the tank in which occupants of a round chariot could speedily move while firing through slits in its sides. Other devices included a submarine, a machine gun, artillery that would obliterate all life in a one-mile radius, and a burning mirror that would set enemy ships alight.

Prompted by his fascination with astronomy, Napier also investigated the application of mathematics to a number of his other obsessions, including even his abhorrence of Catholicism. The first fruit of this was *A Plaine Discovery of the Revelation of St John*, published in 1593. Applying mathematical principles to numbers scattered through the text of the Book of Revelation,

Napier claimed to reveal hidden meanings, including the exact year of the Apocalypse and the conclusion that the pope was the antichrist.

Although Napier considered *A Plaine Discovery* to be his greatest achievement, it was in fact his realisation of the potential of logarithmic numbers that placed him on a directly descending line of discovery from Archimedes to Newton.

Napier treated mathematics as only a hobby that he squeezed into spare time between working on theology. Complaining that he had insufficient time to undertake the necessary calculations of very large numbers and trigonometric functions needed in astronomy, he searched for an easy solution to his mathematical woes. The Eureka moment came when Napier realised that all numbers could be expressed in exponential form and that he could greatly simplify multiplication and division tasks if he added together or subtracted these exponentials.

At its simplest this meant that, for instance, 16 could be written as 2^4 (which means two multiplied by itself four times, or $2 \times 2 \times 2 \times 2$) and 64 as 2^6 (or $2 \times 2 \times 2 \times 2 \times 2 \times 2$). And therefore multiplying 16 by 64 can be easily undertaken by adding the exponentials, so that $2^4 + 2^6 = 2^{10}$, which is 1,024.

Having made his discovery, Napier spent twenty years working out exponential expressions for every number, including the trigonometric functions used in astronomy. For instance, while four can be written as 2^2 and eight as 2^3, the numbers five, six and seven can be written as 2 to the power of a fraction between two and three. Calculating these fractional exponentials was a tedious and arduous task, but by 1614, Napier was ready to publish *Mirifici logarithmorum canonis descriptio* (*Description of the Marvelous Canon of Logarithms*). The book contained thirty-seven pages of explanations, followed by ninety pages of logarithm tables.

Although no more than page upon page of numbers, Napier's tables were immediately seized upon. Their impact on contemporary science was no less than the impact of computers on the science of the twentieth century. By making calculations by hand much easier and quicker, Napier opened the way to many later scientific advances. Without logarithms, it is entirely possible that Kepler and Newton might not have been ideally placed to make their great discoveries.

Napier went on to introduce decimal notation for fractions (an inevitable result of his logarithm calculations) as well as a mnemonic for formulae used in solving spherical triangle calculations. Then, three years after publishing *Mirifici logarithmorum*, Napier unveiled his invention of a mechanical method for simplifying calculations using his logarithm tables. Writing in *Rabdologiae*, a short treatise on how to perform multiplication and division calculations with logarithms, Napier described how to use his 'numbering rods' by lining them up beside each other and reading off the result. Made of ivory, the rods looked quite anatomical and soon became known as Napier's bones. And in the appendix lay another curiosity, a description of a mechanical calculation method that involved metal plates and which appears to be the earliest attempt at a calculating machine.

* * *

Although Napier's bones could be used for calculating square roots and cube roots, they weren't as straightforward to use as a slide-rule, which was invented a few years later by William Oughtred, a minister born at Eton College, where his father taught mathematics.

Obsessed by mathematics to such a degree that he would often sleep only two or three hours a night in order to have more time to pursue his passion in his spare time, Oughtred was determined to devise an instrument that would perform complex multiplication by exploiting Napier's logarithm tables.

In 1621, while Harvey was undertaking the hundreds of dissections that would lead to his discovery of the circulation of blood, Oughtred realised that if he lined up two rulers marked with logarithm scales, he could manipulate them against each other to read off the results of multiplication and division calculations. This first slide-rule was circular, but he went on to develop a straight slide-rule much like the instrument carried and used by generations of scientists, mathematicians and engineers until it was usurped by the electronic pocket calculator in the late 1970s.

Ten years later, Oughtred introduced the multiplication symbol ($\times$) and the abbreviations 'sin' and 'cos' for the sine and cosine trigonometric functions in *Clavis mathematicae* (*The Key to Mathematics*), a book that became a classic and was used by Newton and many other leading mathematicians.

* * *

As a result of scientific investigation and discovery, Britain, like the rest of Western Europe, was by the early decades of the seventeenth century casting off the belief that superstition and religious dogma provided the best explanations of nature.

Between them, Gilbert, Harvey and Napier were laying the foundations of the scientific method. Gilbert's methodical approach to determining the properties and behaviour of magnets had brought rationalism to bear on forces of attraction, such as magnetism and static electricity, which had previously appeared to be mystical and mysterious. William Harvey's anatomical investigations had provided a first step in dispelling long-held Galenic beliefs that the human body was governed by humours, astrological forces and the action of fire and water. Napier's logarithms were dragging mathematics from the limitations of a simple static world of counting and measuring towards the higher mathematics of calculus.

However, it wasn't only the content of the three men's discoveries that was groundbreaking. If anything, their methodology was even more significant. Gilbert's assertion that nothing could be taken for granted if it had not been empirically achieved, Harvey's insistence that theory did not become fact unless it withstood experimental repetition and Napier's development of a method that allowed complex calculations to become a central part of scientific investigation were all radical steps forward.

However, the biggest progression came from one of the most colourful characters of the late Elizabethan period, an eloquent and vociferous promoter of induction and the scientific method. A man who, ironically, was killed by one of the few scientific experiments he ever performed.

Francis Bacon was one of the most colourful and intriguing characters not only in seventeenth-century science, but in the whole of English society. A lifelong consumer of opiates and a connoisseur of weekly rhubarb colonic purges (he was also a patient of William Harvey), Bacon was described by the poet and satirist Alexander Pope as 'the wisest, brightest and meanest of mankind'.

The wisdom and intelligence of which Pope spoke can be clearly seen in Bacon's many writings, the highlight being *Novum organum* (*New Instrument*), which demolished for good the Aristotelian principle of reasoning by deduction. As a result of his influence at the heart of English court society, Bacon's *Novum organum* did more to make experimental science respectable (and even fashionable) among English aristocrats and intellectuals than any writer or book before it.

As for Pope's reference to meanness, Bacon had it in spades. He was nakedly ambitious, highly obsequious to people in authority, and very willing to do the dirty work of others if it bought him preferment. Ostentatious, in love with ceremony, and with an extraordinary ability always to pick the winning side, only to switch allegiance at precisely the moment that his champion fell out of favour, Bacon was unsurprisingly described as cold, arrogant, extravagant and corrupt. Sir Simonds d'Ewes, a contemporary parliamentarian of Bacon, summed up the man best in his diaries, writing that Bacon could have been a good scholar and decent lawyer, but his ambition and pride led to corruption.

Not related to his thirteenth-century namesake, Francis Bacon was born to a prominent family and educated at home before he entered Trinity College, Cambridge, to study law. At Cambridge, Bacon first came into contact with Aristotelian philosophy, which in spite of his admiration of Aristotle as a person he soon rejected as shallow and fundamentally wrong in its objectives.

After Cambridge, Bacon studied at the University of Poitiers, then entered legal practice at Gray's Inn in 1576. A period abroad, staying with the English ambassador to Paris, gave him an insight into diplomacy, the workings of government and the monarchy, but it was brought short by the sudden death of his father in 1579. Left very little in his father's will, Bacon was forced to take his legal career seriously in order to pay off debts, and was soon admitted to the bar. Within five years of his father's death, Bacon had entered Parliament and, aided by his powerful uncle, made rapid progress at the bar. He also wrote several influential tracts on philosophical and religious issues, but his period of greatest influence began when he became acquainted with Robert Devereux, the 2nd Earl of Essex.

A military hero, Essex was the favourite courtier of Elizabeth I. Bacon cultivated the friendship cleverly, soon becoming a confidential aide to the earl, who proposed Bacon as Attorney-General when the position fell vacant in 1594. When Essex's influence proved insufficient, the earl bought Bacon the consolation prize of a property at Twickenham, but it wasn't enough to ensure Bacon's loyalty. A few years later, sensing that Essex was falling out of favour with the queen, Bacon severed his ties with characteristic ruthlessness. Bacon's standing in the queen's eyes immediately improved and his turning on his old friend came full circle when, in 1600, Bacon was appointed to investigate Essex for treachery. Bacon didn't show his former friend any mercy. Essex was convicted and executed in 1601.

Having already cultivated a relationship with James VI of Scotland, Bacon's star rose higher when James succeeded Elizabeth to the English throne as James I in 1603. Bacon was soon knighted. He then made the perceptive move of writing a tract, *Apologie* (*Defence*), to justify his condemnation of Essex, who had supported James's succession. Searching for allies among the new royal courtiers, Bacon found a direct equivalent of Essex in the Duke of Buckingham, James's favourite, and set about nurturing his association with him. This manoeuvring put Bacon into the position of being an unquestioning supporter of the king, a role that did his further rise no harm. In 1607 he became Solicitor General, then Attorney-General in 1613 and Lord Chancellor in 1618, when he was raised to the peerage as Baron Verulam as well as delivering the death sentence at the trial of Sir Walter Raleigh.

After climbing so far and so fast, it was perhaps inevitable that hubris would get the better of Bacon and in 1621 came the fall when, shortly after being made Viscount of St Albans, he became the victim of a political campaign against Buckingham by opponents of the king. An arcane dispute over patent and monopoly legislation lost Bacon the support of James. With the king's protection removed, Buckingham's enemies could strike. A Parliamentary Committee instigated an investigation into allegations that Bacon had accepted two bribes in his capacity as judge. Realising the evidence against him was overwhelming and finding no support from Buckingham, Bacon had no choice but to admit he had taken gifts. His only defence was that the payments had not influenced his decisions, which had gone against the bribers.

The admission did Bacon no favours, merely highlighting instead his duplicity and lack of morals. When the committee went on to cite twenty-eight further cases accusing Bacon or his servants of bribery, Bacon realised that although some of the charges were dubious, he had run out of options. Admitting guilt, he was sentenced to a fine of £40,000 and indefinite imprisonment, but thanks to the king's intervention he escaped payment of the penalty and served only three days in the Tower of London. But, barred from holding any office of state or sitting in Parliament, Bacon's political and court careers were over.

Politics' loss was science's gain. In early life, Bacon had set himself the three goals of serving his country, serving his church and discovering truth. Throughout his court and political career he had devoted time to philosophy, his motivations clearly defined in the phrase he coined in 1597, *scientia potentia est*, loosely translated as 'knowledge is power'.

In 1605, Bacon published *The Proficience and Advancement of Learning*, an encyclopaedia of what was not yet known and a manifesto for how that knowledge might be acquired. In it, Bacon condemned magic, spiritualism and tradition as stultifying and holding back the acquisition of knowledge. The purpose of science, he wrote, was not to support religious doctrine, but to improve human existence, and it should therefore focus only on the empirical world – that is, the world apparent to the senses.

'Men have sought to make a world from their own conception and to draw from their own minds all the material which they employed,' he wrote, 'but if, instead of doing so, they had consulted experience and observation, they would have the facts and not opinions to reason about, and might have ultimately arrived at the knowledge of the laws which govern the material world.'

Unfortunately for Bacon, *Advancement of Learning* was published shortly after the Gunpowder Plot and in the febrile atmosphere of the time it was largely overlooked.

Bacon's most influential publication arrived fifteen years later in 1620. In *Novum organum* Bacon argued that although the Aristotelian principle of reasoning by deduction might be suitable for mathematics, it had no place in science. Instead the laws of science had to be *induced*, that is established as generalisations that were obtained after making a vast number of specific observations. Nature must be 'put to the torture', and made to yield its reluctant secrets to the astute investigator, Bacon said.

In an elegant simile, Bacon explained his recipe for rational enquiry in the form of the behaviour of ants, bees and spiders.

Comparing scholastic natural philosophers to spiders, Bacon said they spun elaborate webs that floated in the air, but all of it was derived from material produced within their own bodies. By contrast, empirics (such as Vesalius with his discovery of valves) were like ants, busily collecting material together but never producing anything from it.

Scientists, Bacon wrote, should be like bees: not only gathering new material like ants, but also digesting and transforming it. A scientist should first conduct experiments in order to amass data and then interpret that data to develop theories that would allow understanding, explanation and prediction from the observations. Furthermore, scientists should strive to disprove their hypotheses by experiment as much as to prove them. Knowledge existed not only in finding evidence to support a hypothesis, but also in conclusively ruling it out.

At the heart of *Novum organum*, Bacon proposed the most fully realised version of the scientific method to date, which he had used to determine that rays emitted by the Sun and by flames transmitted heat, whereas rays from the Moon and stars did not. Now known as the Baconian method, it involved three active steps to determine the cause of a phenomenon. First came a description of facts. Second, each of those facts was classified, ranked and tabulated into three categories – instances in which the characteristic under investigation was observed; instances in which it was absent; and instances in which it was present in varying degrees. In the third step, the tabulated facts were analysed by a process of elimination and inductive reasoning to find the cause of the phenomenon.

For instance, if a football team was successful when managed by a particular manager, not successful when not managed by the same manager, and more or less successful according to the degree of involvement of the manager, then it would be scientifically valid under Bacon's reasoning to say that being managed by the particular manager was causally related to the team's success.

The First Scientists

In the years after his very sudden exit from public life, Bacon was remarkably productive. In addition to some short essays and two historical biographies, he published *De augmentis scientiarum*, an expanded Latin version of *The Proficience and Advancement of Learning* and various parts of his magnum opus, *Magna instaurato*, including the *Sylva sylvarum*, which mixed scientific experiments, personal observations and analytical discussions on a variety of esoteric topics.

Three years after *Novum organum* was published (and two years after his banishment from public life), Bacon published *The New Atlantis*, a utopian novel in which he envisaged a state-sponsored institution, called Solomon's House, in which the best and brightest citizens conducted scientific experiments according to the Baconian method. This concept of an institution in which scientists collaborated to understand and conquer nature, and to apply their collected knowledge to the betterment of society, would prove highly influential in the mid-seventeenth century, when a group of leading scientists established the Royal Society, one of the first such gatherings of scientific minds anywhere in the world since the days of the *Musaeum* in ancient Alexandria, an institution analogous to a modern university where scholars, musicians, artists and poets would meet under the protection of the muses, the Greek goddesses that inspired the creative process.

Bacon didn't live long enough to experience any notable consequences of his writings, although he did die a martyr to the experimental scientific method he advocated. In March 1626, while passing with the king's physician through Highgate in north London, Bacon observed the piles of snow outside his carriage and, in an inspired leap of imagination, wondered if cold might delay the putrefaction of living tissue. He stopped his carriage immediately, bought a hen, and with his own hands stuffed it with snow. But in the cold weather Bacon was seized with a sudden chill, which turned to bronchitis, and he died at the Earl of Arundel's house nearby on 9 April. According to some accounts, he'd attempted to extend his fading lifespan by eating some of the frozen chicken, which if true, would ironically have made it impossible to determine by Baconian method the exact cause (bronchitis or putrid poultry) of his death.

Of all his ideas, Bacon's fact-gathering method gained the most adherents, particularly in mid-seventeenth-century republican England, which was attracted to the Baconian view that the role of science was to improve humankind's lot.

In his statement that knowledge is power, Bacon was also one of the first philosophers to state that the development of ideas and technologies was a more influential shaper of history than the succession of kings and the battles they won. 'Printing, gunpowder and the compass,' he wrote in *Novum organum*. 'These three have changed the whole face and state of things throughout the world; the first in literature, the second in warfare, the third in navigation; whence have followed innumerable changes, in so much that no empire, no sect, no star seems to have exerted greater power and influence in human affairs than these mechanical discoveries.'

Although Bacon's greatest legacy was to make the scientific method respectable, in retrospect he appears to have been surprisingly ill-informed or wilfully ignorant of what was happening around him. Maybe Bacon's obsession with personal betterment, social climbing and intellectual arrogance blinded him to the several real men of science who were already practising the kind of experimental science that he advocated. By the time that Bacon published *Novum organum*, Gilbert had published *De magnete* and died, Harvey was physician to James I and lecturing on his anatomical discoveries, while in Florence Galileo had completed his best work and been forced into silence after the pope declared Copernicanism heretical. All of these scientists reached their findings entirely through experimentation and mostly by rejecting results they had not gathered empirically. And yet Bacon seemed ignorant of them.

As Bacon's physician, Harvey was fully aware of Bacon's writing and must have been galled by its occasional naivety. Certainly Bacon's dismissal of Copernicus' heliocentric model simply because he could not accept the notion of Earth flying through space angered Harvey, who showed his contempt for Bacon when he commented cuttingly that Bacon wrote of natural philosophy and experimentalism 'like a Lord Chancellor'.

Nearly 400 years later, Bacon is better known outside scientific circles than most of the true scientists around him. It's tempting to see the explanation for this in the fact that Bacon the theorist had much better establishment connections than any of his experimenting peers. Certainly his promotion of experimental science was more influential in the seventeenth century than arguably any other natural philosopher. Cynics might also argue that Bacon's classical education, his membership of the bar, high political office and passive theorising garnered less suspicion from English society than the hands-on 'tinkering' of working scientists and medics. Whatever the reasons, Bacon's efforts had at last ensured that the concept of the experimental scientific method based on empirical investigation was firmly established and reliance on mystical explanations was consigned to the past.

Chapter 3

A SOCIETY FOR SCIENCE

In the early part of the seventeenth century, the arcane world of natural philosophy was not alone in experiencing upheaval. Politically and socially, Britain was going through a momentous period of revolutionary change that came to a head at the start of the English Civil War in 1642. For twenty years, as we toyed with the idea of republicanism and Oliver Cromwell led a Britain liberated from sovereign rule, there was a sense that anything was possible. In this febrile atmosphere, scientists and philosophers gathered together in clubs and societies to discuss the issues of the day and to explore the new 'experimental philosophy'.

As the war progressed, the scene of the early scientists' meetings centred on Oxford, the role of which in the history of science during this period has been compared to fifteenth-century Florence and its role in the history of art: a relatively small provincial city from which a clutch of trail-blazing geniuses suddenly burst out of the blue.

Before the Civil War, Oxford and Cambridge had been stable and prosperous. Their universities, established in the Middle Ages to train the clergy, had seen their roles expanded to educate the lay ruling class, which had replaced the clergy in political and administrative affairs. For most of that period, the classical orthodoxies of Aristotle and Galen had prevailed. Then, just as experimentalism and the Baconian method were gaining support and interest, the war broke out. After the second pitched battle of the war, at Turnham Green in 1642, Charles I was forced to withdraw to Oxford, which became his military headquarters. At the university, student numbers and the teaching system collapsed, and after the defeat of the king in 1648 a Puritan purge of the university further depleted morale and resources.

The collapse of the university sent a swathe of scholars running to London. Before that, whilst still in Oxford, many of them had met informally to discuss natural philosophy and ways of advancing technology. When they arrived in London, these groupings of like-minded Oxford scholars became more formalised. One of the groups was brought together by John Wilkins, a clergyman with an interest in science and one of the founders of a movement seeking to develop a religious philosophy called natural theology that was compatible with modern developments in science.

Called the Invisible College because they didn't have a regular meeting place, the group included, among others, Jonathan Goddard, army surgeon to the Parliamentary forces; George Ent, an anatomist who was among Harvey's major defenders; Francis Glisson, a physician who researched the anatomy of the liver; Samuel Foster, Professor of Astronomy at Gresham College; Theodore Haak, a German scholar at Gresham College; and John Wallis, a mathematician and physicist.

Motivated by utopian aspirations, this group of intellectuals exchanged ideas by correspondence and engaged in promoting science and technology. By the mid-1640s, they had become more organised, meeting weekly, usually at Gresham College, as described by John Wallis.

About the year 1645, while I lived in London at a time when, by our civil wars, academical studies were much interrupted in both our Universities, I had the opportunity of being acquainted with divers worthy persons, inquisitive into natural philosophy, and other parts of human learning; and particularly of what has been called the New Philosophy, or Experimental Philosophy.

We did by agreements, divers of us, meet weekly in London on a certain day, to treat and discourse of such affairs. These meetings we held sometimes at Dr Goddard's lodgings in Wood Street or some convenient place near, on occasion of his keeping an operator in his house for grinding glasses for telescopes and microscopes; sometimes at a convenient place in Cheapside, and sometimes at Gresham College, or some place near adjoining.

Our business was (precluding matters of theology and state affairs), to discourse and consider of Philosophical Enquiries, and such as related thereunto: as physic, anatomy, geometry, astronomy, navigation, statics, magnetics, chemics, mechanics, and natural experiments; with the state of these studies, as then cultivated at home and abroad.

As described by Wallis, their discussions ranged extremely widely:

We then discoursed of the circulation of the blood, the valves in the veins, the venae lactae, the lymphatic vessels, the Copernican hypothesis, the nature of comets and new stars, the satellites of Jupiter, the oval shape (as it then appeared) of Saturn, the spots in the sun, and its turning on its own axis, the inequalities and selenography of the moon, the several phases of Venus and Mercury, the improvement of telescopes, and grinding of glasses for that purpose, the weight of air, the possibility, or impossibility of vacuities, and nature's abhorrence thereof, the Torricellian experiment in quicksilver, the descent of heavy bodies, and the degrees of acceleration therein; and divers other things of like nature. Some of which were then but new discoveries, and others not so generally known and embraced, as now they are, with other things appertaining to what has been called The New Philosophy, which from the times of Galileo at Florence, and Sir Francis Bacon (Lord Verulam) in England, has been much cultivated in Italy, France, Germany, and other parts abroad, as well as with us in England.

Judging by Wallis's description, the Invisible College was an association with high ambitions, but it started to disperse in 1648, when parliamentary commissioners appointed Wilkins to the post of warden of Wadham College.

Wilkins began his tenure at Wadham with the aim of fostering a climate of political and religious tolerance and soon attracted talented minds back to Oxford, including Goddard. Almost immediately, Wilkins formed a salon at Oxford based on the Invisible College, but adding several new members from among the latest generation of students at the university and from a band of 'virtuosi' polymaths who had been trapped at Oxford during the Civil War. They included Christopher Wren, a young protégé at that time more celebrated for removing the spleens of dogs than for building churches; the philosopher John Locke; William Petty, an anatomist and demographer; Thomas Willis, an upwardly mobile farmer's son who had studied medicine under Harvey; Seth Ward, who investigated astronomy and mathematics; and later in 1653, when he arrived at Christ Church, Robert Hooke, who coined the term 'cell'. Hooke would later become the greatest experimentalist of his generation but he joined the group as an assistant to Willis initially before being employed by Robert Boyle, an aristocratic Irish gentleman scientist and philosopher who moved to Oxford in 1654.

> During the bloody civil war, the king and his court took refuge in Oxford from the fighting. Among the courtiers was a young man called Christopher Wren. Today we know him as an architect for the magnificent buildings he put up in London and Oxford.
>
> But there was so much more to Christopher Wren. The civil war had turned Wren's life upside down. It left Wren with a desperate yearning for certainty, stability and truth.
>
> In Oxford, Wren's academic brilliance had been spotted by John Wilkins, a cleric who ran a club for experimental philosophers. These men were passionate in their pursuit of knowledge and their desire to understand the natural world through reason, logic and experiment rather than through mysticism. Wren had found his niche.
>
> David Attenborough

Meeting every Wednesday afternoon at Wadham, the group acquired a new name – the Oxford Experimental Philosophy Club – and set about debunking Aristotle and Galen by performing wide-ranging experiments. Thanks to Wilkins' skills in creating a safe haven away from the ideological, political and religious ructions of the Civil War, Oxford became a hotbed of scientific experimentation and intellectual enquiry. Their activities included the creation of a universal language, the microscopic description of insects and the telescopic examination of the Moon.

In 1654, the diarist John Evelyn, a contemporary of Pepys, visited Oxford, met many members of the Experimental Philosophy Club, heard their disputations, visited the Anatomy School and the Physic Garden, and concluded that this 'is doubtless the leading university now in the whole world'.

This concentration in Oxford of a generation of self-styled 'virtuosi' scientists, most of them polymaths who brought philosophies and knowledge from a wide area to bear on their scientific experimentation, was hugely significant in the development of English science. A few years later it would lead to the development of the Royal Society in London, but before then it fostered the scientific careers of a number of the leading experimentalists of their day.

* * *

Among those experimentalists was Thomas Willis, a short, stammering, red-haired country boy whose life was profoundly shaped by the Civil War. Intending to pursue a career in the Church, Willis arrived at Oxford in the late 1630s, achieving his Bachelor of Arts in 1639 and a Masters in 1642, shortly before war broke out. A Royalist and an Anglican, Willis abandoned theology to fight for the king, serving for two years in an auxiliary regiment of university members raised by the Earl of Dover to defend Oxford.

Partly as a reward for his loyalty to Charles I, Willis was recommended in 1646 for a medical degree by a fellow Royalist, Thomas Clayton, the Regius Professor of Medicine. His Bachelor of Medicine was awarded only weeks before the Parliamentary takeover of Oxford. Although it came at the last possible moment, the degree immediately served Willis well, sparing him from the persecution that Parliamentary forces exerted on other theologians at the university.

Ironically, the Civil War also shaped Willis's original mind. At that time, medical training at Oxford lasted fourteen years, mostly centred on study of the classical works of anatomy and medicine. However, with the university's usual procedures thrown into turmoil by war, Willis escaped the stifling fate of reading Aristotle and Galen. Liberated from an adherence to the orthodoxy of the past, he pursued an experimental approach to medicine that led him to groundbreaking discoveries in neurology, including linking emotions and several illnesses to the functioning of the brain.

Often overlooked in the pantheon of seventeenth-century heroes of science, Willis's discoveries completely changed our understanding of the functioning of the brain. Before Willis, the brain was thought to be relatively primitive and, in the words of the philosopher Henry Moore, no more capable of thought than 'a bowl of curds'. To some extent this was understandable. In an era long before the advent of refrigeration and sophisticated chemical preservatives, it was often difficult to determine the physiology and function of organs after death. Muscular organs, such as the heart, remained largely intact for some time. Other organs, such as the liver, deteriorated slowly. However, the brain rapidly turned to a soft mush.

This sudden decomposition of the brain into a custardy porridge led early natural philosophers to consider it to be a secondary organ. Compared with the heart, which they felt pounding faster in their own chests when they became animated, the brain felt lifeless. Consequently, the heart was deemed to be the seat of the soul and the brain merely an organ for cooling spirit from the heart. According to Aristotle, it was only one of three places in the body in which a type of soul resided – lust in the gut, emotion in the heart, and rationality in the head – and relics of this separation are still seen in everyday speech that ascribe to the heart or bowels emotions that we now know originate in the head. In distress, for instance, we talk about grieving from the heart. When in love, we symbolise our feelings with heart shapes. When love is over, we talk about being heartbroken, and if things go wrong for us, we say we're gutted. And in situations requiring courage, we talk of showing guts.

Early anatomists such as Galen discovered very little to challenge Aristotle's view. Having observed the effects of brain injuries on mental processes and linked nerve functions to the brain, Galen correctly attributed responsibility for sensation and thought to it. But he never attempted a dissection of the head and declared the brain to be a cold, moist organ (made of sperm) that distributed the four humour fluids through the nerves to other parts of the body.

Anyone who challenged the Galenic interpretation could be accused of committing heresy, but that all changed when Willis, inspired by Harvey's discovery of the circulation of blood, decided to trace the course of blood further through the body.

Using Harvey's experimental methods of dissection and experimental testing, Willis tracked the flow of blood to the brain, but only after a long and frequently unpleasant apprenticeship in general medicine and dissection.

After receiving his Oxford degree in 1646, Willis set up a modest medical practice in his Christ Church rooms. On market days he travelled to surrounding villages such as Abingdon, where he operated as an itinerant dispenser of medical advice. Like many of his contemporaries, Willis operated as a 'pisse-prophet', diagnosing disease by inspecting the colour of his patients' urine. Willis would sit rather majestically in the town square while friends of the bedridden proffered samples and, when requested by Willis, they would also supply a report on the taste of their own effluent. For any imbalance in humours, he prescribed time-honoured but dubious remedies such as ground-up millipedes or amulets of mistletoe. Patients with migraines would have holes drilled into their heads, while sufferers of other ailments were subjected to transfusions of lamb's blood or remedies made from cut-up puppies.

Even while practising primitive Galenic medicine, Willis maintained an open and inquiring mind. Having taken several of his patients' urine samples home, he noticed that flies were attracted to some samples more than others. He suggested the cause was the sweetness of the urine, which he found distinguished one form of polyuria (excessive urination) from other causes of the symptom, such as chronic kidney disease. As a result of this, doctors still refer to diabetes subtypes as mellitus (meaning sugar diabetes) and insipidus (water diabetes).

In his spare time Willis worked in the laboratory, researching iatrochemistry, which involved finding naturally occurring chemicals to treat diseases. Working with Robert Hooke, Robert Boyle and other members of the Oxford Experimental Philosophy Club, Willis contributed to the expense of operating John Wilkins' chemical laboratory at Wadham College and soon gained a reputation as one of the best chemists in the group.

In 1649, Willis's enlightenment from the quackery of Galenic medicine began in earnest with the arrival at Oxford of William Petty, a young physician. Having studied at Leiden in Holland and Paris, Petty subscribed to Parisian theories that the body was in essence a mechanical contraption that could be understood by meticulous dissection.

Willis's fascination with the human brain began paradoxically in the year that Charles I lost his head. At first he assisted and watched Petty, who as Tomlins Reader in Anatomy at Oxford was granted permission by a royal charter of 1636 to claim the body of anyone executed within 21 miles of Oxford. Intended to increase the availability of cadavers after an increase in medical student numbers had led to a chronic shortage of bodies for the students' requisite two dissections each, the charter gave Petty and Willis a rich source of research material. Performing dissections together at Petty's residence in Bulkley Hall – a building off High Street in Oxford that is now the site of a Thai restaurant – they opened up dozens of executed criminals to peer at their inner workings.

When Willis had seen and learnt enough, he decided to have a go himself. Assisted by Richard Lower, who later inherited Willis's medical practice, and equipped with a knowledge of anatomy from Petty, Willis soon turned his attentions to the brain. 'I addicted myself to the opening of heads,' he told his colleagues at the Oxford Experimental Philosophy Club, with the hope that he would, as he put it, 'unlock the secret places of man's mind'.

Willis worked with Christopher Wren, John Locke and Ralph Bathurst, who like Willis was an Oxford theologian who turned to medicine in the Civil War. They dissected living dogs to explore the beating of the heart; they investigated the consequences of removing various internal organs, some of which, astonishingly, the dogs survived; and they experimented with the possibility of transfusing blood.

The strangest event in this somewhat macabre period occurred on 14 December 1650, when Wallis, now assisted by Petty, resuscitated an apparently dead 22-year-old woman who he was preparing to dissect after she had been hanged on the Cattle Yard in Oxford.

Anne Green had been employed as a housemaid at a house in Oxfordshire. Seduced by her employer's grandson, she became pregnant. She hid her pregnancy and gave birth to a premature child secretly. However, the dead body of the new-born baby was found and Anne Green was arrested. Convicted of murdering her child, she was sentenced to die by the usual method: hanging.

The execution was as brutal as any carried out at that time. Anne Green climbed a ladder to the gallows, a rope was laid around her neck and she was pushed off the ladder. Onlookers pinched her breasts and hung on her legs, lifting up her body, then jerking it downwards with all their weight to ensure she was dead. Fearing that the rope might break, a court usher urged the spectators to leave the body alone until, after hanging for half an hour, Anne Green was pronounced dead, cut down and placed in a coffin.

As per usual, Willis received the body, but when he opened the coffin to begin dissection at Petty's home, the apparently dead woman emitted a strange noise from her throat, then took a breath. Startled, Willis and Petty immediately raised Anne Green to an upright position in her coffin and rubbed her extremities. To provoke more coughing, they poured hot cordial into her mouth and tickled her throat with a feather. Anne Green opened her eyes.

Willis bled Green of five ounces of blood, and swaddled and wrapped her limbs in hot bandages to force blood to her brain, then placed her in a bed beside another woman to keep her warm. Twelve hours later, Anne Green could speak. The next day she could respond to questions. After four days, she ate solid food. Two weeks later, her memory returned and she could remember a man wearing felt, most likely the executioner, standing behind her. She recovered fully within a month, going on to move to the countryside (taking with her the coffin in which she had lain), where she married and bore three more children.

Having kept copious notes on their resuscitation of Anne Green (including pulse, respiration rate, blood pressure, visual observations of her state and the return of her memory), Willis and Petty reported on their findings. Anne Green's memory, they said, behaved like a 'clock which has had the weight removed and later hanged on again, whereby the clock starts'.

The incident did nothing to improve executions. Only in the second half of the nineteenth century was a fall to break the neck introduced to hangings. Before then, when children were executed, their parents would often cling to them in the hope that their additional weight would help them die quicker.

However, Anne Green's return from the dead did wonders for Willis. His medical practice immediately prospered, he married the daughter of the Dean of Christ Church, and in 1657 moved to Beam Hall, on Merton Street opposite Merton College, where he employed an apothecary and five servants.

At Beam Hall, Willis devoted himself to dissecting human brains, usually with Petty, but often also with Wren and Millington present 'to confer and reason about the uses of the parts'. Robert Boyle, a young and ambitious son of an Irish aristocrat who had recently joined the Experimental Philosophy Club, had discovered that immersing organs in pure alcohol delayed their putrefaction, allowing Willis to examine organs and samples at his leisure.

Instead of the traditional method of dissecting the brain *in situ* from above, Willis approached it from below and removed it from the skull. In some cases, Wren injected animals intravenously with dyes shortly before death so that he could follow the path of blood through the brain. Then, when the brain had been removed and preserved in alcohol, Willis sliced it from the base upwards, passing the samples to Wren, who examined them under microscopes he had designed with Robert Hooke.

Without the benefit of several advances in scientific techniques developed by the Oxford circle, to which he was deeply indebted, Willis might not have achieved his remarkable findings such as his discovery of the ring of arteries at the base of the brain, known to this day as Willis's circle.

Through Wren's dye injections, Willis discovered that blockage of only one of the four main cerebral arteries would not lead to apoplexy. He also examined the network of blood vessels that covered the brain, giving it the appearance of a 'curious quilted ball'.

In 1660, Willis was appointed the Sedleian Professor of Natural Philosophy in Oxford by Gilbert Sheldon, who later became Archbishop of Canterbury and Chancellor of the University of Oxford. Soon after becoming Sedleian professor, Willis's interests shifted from iatrochemistry and tracing blood vessels to the functions of the brain: neuroanatomy and neurology.

Through painstaking dissections and observations, Willis traced the intricate network of nerves emanating from the brain to control the various organs. By constricting nerves (a procedure called ligation) in dogs, he traced the functions of the cranial nerves. Then, following nerves down the spine to show that the cerebellum controlled the beating of the heart, Willis became the first to prove that the brain, not the heart, was the prime organ.

With that proved, it followed that all the functions attributed to the soul such as intellect, passion and memory would be found in the brain.

Willis also investigated the fluid-filled ventricles at the centre of the brain. According to Galen, these were essential and contained an 'animal spirit' distilled from sensory signals, reason, thought, judgement and memory to produce the essence of all intellect. But Willis discovered they were actually no more than a 'complication of the brain infoldings'.

In 1664, Willis published the results of his findings in *Cerebri anatome* (*Anatomy of the Brain*), a book that was unsurpassed in its contribution to neuroanatomy for almost 200 years. Establishing the concept of neurology, Willis made the bold claim that the brain was 'the chief seat of the Rational Soul in a man ... and as the chief mover in the animal Machine'. With beautiful illustrations by Wren that portray the brain as a delicate, complex organ, the book followed the same format as William Harvey's *De motu cordis*: intricate diagrams, comparisons between animal and human anatomies, and supporting empirical evidence, in this case patient case histories. Two-thirds of the book addressed the brain. The remainder described the cranial, spinal and autonomic nerves. *Cerebri anatome* was extremely successful, four Latin editions appearing in 1664 alone, so it was perhaps understandable that Willis, having written the first book that ascribed emotion and intellect to the brain and the nervous system, a potentially heretical assertion, went to the trouble of including a lengthy and effusive dedication to Sheldon, who by the time of publication had become the Archbishop of Canterbury. Insisting that his findings did not make him an atheist and that they would be a service to Christianity, he wrote that nature was not only a machine, but also a 'table of the Divine Word, and the greater Bible' and that the brain within nature was the 'Chapel of the Deity'.

Willis continued his practice of dedication to the archbishop in his next two books, *Pathologiae cerebri, et nervosi generis specimen* (*A Model of a Kind of Cerebral and Nervous Pathology*), published in 1667, and *De anima brutorum quae homine vitalis ac sensitiva est* (*Two Discourses Concerning the Soul of Brutes*), published five years later. In these books, Willis laid further foundation stones of contemporary neuroscience, in particular the assertion that many afflictions, both neurological and psychological, can be cured by 'manipulating the atoms which compose it'.

Before Willis proved otherwise, many psychiatric conditions and physiological states, such as mania, melancholia, insomnia, lethargy, narcosis and hysteria had been associated with disorders of the lungs, spleen or uterus. From his case notes and careful observation of healthy, ill and disabled patients, Willis linked the conditions to brain disorders.

As a result of his publications and the reputation he gained after resuscitating Anne Petty, Willis became one of the most trusted physicians in Oxford, expanding into a dilapidated coaching house on the High Street near Magdalen Bridge in order to cater to rich travellers. By providing consultations, lodging, nursing care, proprietary medicines and spa waters, Willis earned enough to buy himself a Herefordshire estate. With capital of £300 and annual earnings of the same amount, he had the largest income in Oxford by the time he departed Beam Hall for London in 1667. When he arrived in the capital, Willis opened a practice that became the most fashionable and profitable of the period, charging wealthy patients the highest fees for his services, but treating the poor for free until he died in 1675.

* * *

By the time Willis moved to London in 1667, Britain had become a very different place. The realities of anti-monarchism had failed to live up to the promise, the Parliamentary republic had collapsed and Charles II was back on the throne after returning from exile in France in May 1660.

In the November of the year that Charles II returned, immediately after a lecture at Gresham College by Christopher Wren, then Gresham Professor of Anatomy, twelve men decided to establish an official society 'for the promoting of Physio-Mathematical Experimental Learning'. In many ways, the founding of such a body was a natural metamorphosis of the Invisible College and the Oxford Experimental Philosophy Club of the Civil War years. Several of the founding members, such as Robert Boyle and John Wilkins, William Petty and Christopher Wren, had been members of one of those loose affiliations of like-minded experimentalists, and their aims were broadly as before: to promote experimental science. However, the twelve men who met at Gresham College that night had much loftier ambitions: to become a truly national society, open to anyone with an interest in science. From the outset, they wanted the society to combine the roles of a research institute that held weekly meetings to witness experiments, a clearing-house for knowledge, and a forum for arbitration and discussion.

Of the dozen founding members, William Brouncker, an Irish aristocrat mathematician, had the closest association with the king and was consequently tasked with sounding out the monarch.

Assured by Lord Brouncker that Charles would be favourable to seeing such an enterprise succeed, they resolved to seek a royal charter. Almost immediately, the king agreed to support (and later act as patron to) a society of his leading thinkers, many of whom had been sworn enemies during the Civil War, but who shared the common assets of inquiring minds and a grasp of science.

The founding members compiled a list of forty others to invite to participate in the venture. Sir Robert Moray, a Scottish former spy and soldier with renowned diplomatic skills, was chosen to bring together a group of men who had seemingly irreconcilable differences from the Parliamentary–Royalist divide. The result was the Royal Society of London for the Promotion of Natural Knowledge. Better known simply as the Royal Society, it was composed largely of Puritan sympathisers and adherents of the new experimentalism. Consequently, it received little more than moral support from the crown, unlike academies on the European continent, which were established by the state and whose members gained an income. However, unlike its foreign counterparts, its maverick status enabled the Royal Society to maintain independence from religious or regal influence.

With a royal charter granted in July 1662, the Society named Lord Brouncker as its first president and chose *Nullius in Verba* – Nothing in Words – as its motto, indicating that it believed that experimental deeds spoke louder than scholastic words. John Wilkins became the first of the Society's Secretaries for biological sciences and Robert Hooke was appointed as Curator of Experiments.

A SOCIETY FOR SCIENCE

> The Royal Society's interest and global reach rapidly widened. It soon
> became the repository for scientific information from all over the
> world. It received data about tides from Jesuit priests in China, it
> commissioned reports from Barbary pirates, and geographies from the
> new colonies in India and America. It was the place to be if you
> wanted to enhance your scientific reputation, and Robert Hooke as
> director of experiments was right at the heart of it.
>
> Richard Dawkins

The royal charter enabled the society to publish, and it went on to produce many key titles in science. In March 1665, Henry Oldenburg, the Society's first secretary for physical sciences, began publishing the first (and what has become the longest-standing) scientific journal, the *Philosophical Transactions: giving some Accompt of the Present Undertakings, Studies, and Labours, of the Ingenious in many Considerable Parts of the World*, now known simply as the *Philosophical Transactions*. In publishing the journal, Oldenburg established one of the main procedures of modern science, namely subjecting an author's scholarly work, research, hypothesis and conclusions to the scrutiny of others who are experts in the same field.

Oldenburg's intention was that the *Philosophical Transactions* could be used to establish precedence so that the first discoverers of a phenomenon could be publicly acknowledged. From the outset, he did not publish all the material he received. Instead the Council of the Royal Society reviewed the contributions before approving a selection for publication. Although primitive, it was the first recorded instance of peer review. The four functions set up by Oldenburg for *Philosophical Transactions* (registration, dissemination, peer review and archival record) were soon recognised as so fundamental to the way that science was carried out that all subsequent journals, even those published electronically in the twenty-first century, conformed to Oldenburg's model.

* * *

One of the first scientists to publish his research in the *Philosophical Transactions* was a shy polymath who was heavily influenced by Bacon's writing. A leading founder of the Royal Society and the main contender for the title of father of modern chemistry,

Robert Boyle could also be regarded as the first true experimentalist. Although Harvey and Willis were men of science, in many ways they were as much discoverers as experimentalists: they explored the human body and unearthed its workings, using experiments to verify their discoveries. Boyle had a quite different approach: he started from a position of theoretical consideration of the physical world, albeit one that was rooted in alchemy, then devised controlled experiments to test his hypotheses. In the process, Boyle developed a mechanistic view of nature to displace Aristotelian scholasticism and became a leading figure in the new science, ruminating on its rationale, implications and relationship with religion.

Born in 1627, Boyle was the youngest of seven sons and fourteen children of the first Earl of Cork, an adventurous and highly ambitious entrepreneur who arrived almost penniless in Ireland but who, in rising to Lord High Treasurer, became the richest and one of the most powerful men in Britain.

Although his father had been ennobled only twenty-four years earlier, when he married Catherine, the daughter of the Secretary of State for Ireland, Robert had a typically aristocratic upbringing, looked after by an Irish nurse who, instead of using a cradle, laid him in a pendulous satchel with a slit for his head to stick out.

At the age of eight, by which time he had learnt Latin, Greek and French at home, Boyle was sent with his brother Francis to Eton College. The earl, who was not an easy man, wanted the two youngest Boyle boys to be toughened up after the death of their mother, the earl's second wife.

After three years at Eton, where Robert's studiousness and sickly, pale appearance stood out, the Boyle brothers embarked on a continental tour led by Isaac Marcombes, a Huguenot intellectual. A few months into the trip they reached Geneva, where Marcombes schooled them and where the 14-year-old Boyle experienced a major thunderstorm that changed his life. Thinking the end of the world had arrived, he vowed that if he survived the night he would spend the remainder of his days 'more religiously and watchfully employed'.

Boyle's new-found drive was soon apparent. Fascinated with Galileo, Boyle became determined to meet the great Italian physicist and astronomer; an ambition he thought he might achieve when in 1641, after two years in Geneva, he set out with his brother and Marcombes for Florence. But they arrived shortly before Galileo died, which served only to increase Boyle's curiosity about the leading scientist of his era. The party wintered in Florence, taking the opportunity to study the 'new paradoxes of the great star-gazer', as Boyle described it in his autobiography. In the spring of 1642, they moved on to Rome, returning to England in 1644 to take up residence at Stalbridge, the Dorset estate left to Boyle by his father.

Ensconced at Stalbridge, Boyle began his career as an author, initially writing on literary and moral issues. However, his fascination with Galileo had sparked a wider interest in science and by the late 1640s he was travelling regularly to London to take part in the meetings of John Wilkins' Invisible College, which often met at the home of his favourite sister Katherine. While the capital was racked by the upheavals of the Civil War, Boyle was becoming acquainted with the latest developments in natural philosophy, in particular the new vogue for Baconism. Boyle found the concept of experimentalism very appealing; it prompted him to set up a laboratory at Stalbridge in 1649 and to begin conducting experiments. From the summer of that year, he started to discuss scientific topics in his writing, which included experiments in alchemy and early chemistry, as well as his microscopic examination of living things.

Boyle had been groomed by his family to be a landowning member of the Irish aristocracy, but he turned his back on it all, and became obsessed with an obscure cult: science.

For Boyle, science wasn't just an idle philosophical investigation, or a gentlemanly hobby. It was an almost spiritually powerful tool to explore the true nature of everything around us, and as a committed Christian he believed it was a direct route to seeing the true face of God.

Science for Boyle was about asking questions about everything, and you can get a sense of his incredible scientific obsession through some of the titles of his writings: *His Manner of Giving Meat to a Dog*; *Upon the Sight of a Fair Milk Maid Singing to her Cow*; *Upon My Spaniel's Carefulness Not to Lose Me in a Strange Place*; *Upon the Eating of Oysters*.

But for me, Boyle reveals his true scientific colours in a paper he wrote following an unusual encounter in his country home. One night in 1671, a servant woke Boyle with the news that there was something terrifying and alive in the pantry. Boyle crept into the kitchen and found a leg of lamb glowing with a strange light. Two hundred years later this became known as photoluminescence; a glowing bacteria like you might see in the sea. But back then Boyle had no idea what it was.

For an obsessively curious scientist like Boyle, this was just the kind of mystery he loved, and all that night he experimented on the meat. He weighed it, measured its temperature; he magnetised it, sliced it, rubbed it, squeezed it; he placed it between two sheets of glass and he even put it in wine.

It seems to me that he had no clear idea of what he would find. It was a wonderfully thorough, scattergun approach. And this approach defined a new kind of experimentation: one where the scientist became part of the experiment, by actively testing. By prodding and squeezing.

Boyle was one of the first true masters of this new kind of experimental science. And nowadays, we do experiments in exactly the same way. We set up the apparatus and try things out. Some things fail, in fact most things fail, but what's interesting is that you need to observe those failures because through them you begin to understand how you might overcome those problems.

James Dyson

Having fought hard to avoid taking sides in the Civil War (as an aristocrat, his father was a staunch Royalist, but Robert had no particular allegiance or sympathy with either side), Boyle moved to Ireland in 1652 to check on his estates, only to receive huge tracts of land from Cromwell after the Parliamentarian defeat of the Irish in the same year. The apportioning of land to the colonialists made Boyle an extremely rich man, now in the enviable position of being able to devote all his time to science.

The next year, Boyle returned to England and bumped into an old acquaintance from the Invisible College when he was on one of his frequent visits from Oxford to the capital. It was Wilkins, who told Boyle of the Experimental Philosophy Club that he'd instituted at Oxford since taking over as warden of Wadham College five years earlier. Wilkins encouraged Boyle to join them, even offering Boyle accommodation at the college. Encouraged by Wilkins, Boyle moved in 1654 to Oxford, hiring his own rooms on the High Street so that he could set up a personal laboratory of his own design. Meeting frequently with other experimentalists, including Willis and Wren, Boyle's studies became more focused. Meanwhile, he wrote widely on subjects ranging from chemistry and anatomy to philosophy, theology and even religious romances. However, his writing in this period remained unpublished until several years later.

The late 1650s were an intensely active period for the gentleman scientist Boyle. He moved on from his classic experiments, such as showing the reconstitution of saltpetre, to assembling what would in modern science be called a research group and developing his own apparatus for conducting experiments. In both of these he was aided by Robert Hooke, an exceptionally bright, self-educated and mechanically minded young man who had won a chorister's place at Oxford. Boyle employed Hooke as his assistant in 1655, but Hooke's influence stretched far beyond the lab, even extending as far as to help Boyle's understanding of the major natural philosophers of continental Europe, in particular René Descartes, who fused algebra and geometry into his system of Cartesian coordinates. But it was in Boyle's laboratory that Hooke really proved his worth. Boyle was intrigued by William Harvey's discovery of the circulation of blood, which he thought begged as many questions as it answered, and wondered whether there might be some connection between the circulatory system and the air that we breathe.

Boyle wanted to find out exactly what air was; what this strange invisible substance was that was odourless and impossible to feel, yet that surrounded him all the time.

To elucidate the nature of air, Boyle asked Hooke to help him make an air pump similar to the one he'd read that Otto von Guericke, the German physicist, had recently used to dismiss Aristotle's contention that nature abhors a vacuum, a dictum accepted without question by generations of scholastics.

Working together in Boyle's Oxford room, Hooke and Boyle built a device that they called a 'machina Boyleana' and a 'pneumatical engine'. They enlisted the best glassmakers in the country – exiles from Venice – to build a glass observation sphere. They hired gun makers to build the piston used to evacuate the air. They called on the watchmakers of Ludgate Circus to work on the cog system that would lever the piston down effectively. Made of glass and brass, the device took several years to manufacture but when, in 1659, it was ready, it caused a sensation.

The pump that von Guericke had used to demonstrate the power of a vacuum to the emperor Ferdinand III had required the strenuous efforts of two men and had provided inconsistent results. Boyle's pump could be operated easily and efficiently by one man, provided that man was Hooke. Other operators, Boyle included, struggled to achieve consistent results, but in Hooke's hands it was dependable. Its chamber was accessible and it was large enough to carry animate objects, such as a bird or mouse. But the most impressive thing about it was that the air pump opened a window on a new world.

Boyle used his pump to illustrate the characteristics of air by studying the effects of its withdrawal on flame, light, and living creatures. He demonstrated that the sounds of a bell ringing and a clock ticking in the vacuum chamber faded as the air was removed, proving that air was essential for the transmission of sound. Likewise, the flickering and eventual extinguishing of a candle flame as the air was sucked out of the chamber proved that air was necessary for combustion. And when he trapped a mouse in the chamber, he was able to show that air was necessary for life – the mouse slowly expiring as its life support was literally sucked out of it with every stroke of the vacuum pump. By contrast, a snake merely weakened as the air was withdrawn, Boyle surmising that cold-blooded creatures were less dependent on air than their warm-blooded brethren.

By placing a candle in his vacuum chamber and watching it go out, Boyle showed that nothing could burn without air. Even more strangely, lukewarm water spontaneously boiled as he sucked out the air.

He then put a pig's bladder filled with air in the chamber. As he pumped the air out of the chamber, he saw the balloon swell and burst as the pressure around it decreased.

He'd discovered that air isn't just lifeless and inert. It has properties of its own. It has elasticity, weight, pressure and volume.

James Dyson

To Boyle, the vacuum pump experiments confirmed his mechanical view of nature and smashed the Aristotelian, non-empirical approach to science. He believed that these experiments, and others that he did not detail at this time, showed that air had a certain weight and 'spring'.

Having begun to publish in 1659, albeit initially on philosophy and morals, Boyle published the findings of his vacuum pump experiments in 1660 in his first science book, which had the snappy title of *New Experiments Physico-Mechanical, Touching the Spring of the Air and its Effects (Made for the most part, in a New Pneumatical Engine)*. With his large inherited fortune behind him, Boyle could afford to have his books printed at the highest quality and he soon became the best customer of several notable London printers.

> Boyle's air pump was a huge turning point. It demonstrated that there was an invisible world all around us whose laws we could come to understand through reason.
>
> The question was: how far could this same way of thinking apply? Were the movements of the planets and stars also subject to hidden laws?
>
> Stephen Hawking

The next year saw the publication of *The Sceptical Chymist*, the first true chemistry text, a cornerstone in the development of empirical science, and arguably Boyle's most significant book. In *The Sceptical Chymist*, Boyle abandoned the alchemists' tradition of secrecy and published full experimental details, including his unsuccessful experiments, which in the proper traditions of modern science had as much to teach as experiments that ended in success. That said, even Boyle, the so-called father of chemistry, never fully abandoned his faith in some aspects of alchemy, such as transmutation of the elements, remaining convinced until late in his career that he was on the verge of succeeding in turning mercury into gold.

Nevertheless, with *The Sceptical Chymist*, Boyle managed at least to turn alchemy into chemistry and thereby divorced it from its historic links with medicine and pharmacy, establishing chemistry as an independent science. In defining an element as any substance that could not be broken down into simpler substances and which could be identified only by experiment, he broke with Aristotle's classical view that elements were mystical entities such as fire, water, salt and mercury that were necessary ingredients of all bodies. He also showed how elements could be combined into compounds that had quite different properties to their constituent elements and then obtained again by breaking down that compound. And he touched on the idea of atomic structure in his assertion that matter was ultimately composed of 'corpuscles' of varying sizes and types that arranged themselves to form elements.

By the time that *The Sceptical Chymist* was published, Boyle's views on experimentalism and the nature of air expounded in his previous book, *New Experiments ... its Effects*, had attracted as many critics as converts to his cause. Among the critics were the philosopher Thomas Hobbes, the Dutch scholar Anthony Deusing, and Franciscus Linus, a Jesuit priest and natural philosopher, who objected to Boyle's contention that the pressure of the atmosphere (which Boyle called the 'spring of air') was sufficient to raise a mercury column in a barometer by about 29 inches and that the void above the mercury in the closed end of a barometer was a vacuum.

Unwilling to accept that Boyle's statement contradicted Aristotle's maxim that nature abhors a vacuum, Linus argued that something invisible above the mercury must be holding it up.

To address Linus' objection, Boyle had Hooke prepare a J-shaped glass tube. Mercury was poured into the tube, trapping a small amount of gas in the short upright limb. Hooke then performed a series of experiments in which he increased or decreased the volume of the trapped gas by varying the pressure on the long limb of mercury. The experiments led Boyle to the discovery that the pressure and volume of a gas are inversely related. That meant a doubling of pressure would reduce the volume by a half, while a trebling of pressure would reduce volume to a third. Easing off the pressure would allow the gas to expand by an equivalent amount.

This relationship (expressed as $P \times V = k$, where P is pressure, V is volume and k is constant) is now known to every secondary school science student as Boyle's Law, although a fairer attribution would be to call it the Boyle-Hooke Law as Boyle only thought of the experiment, while Hooke worked out how to conduct it and how to make it work.

Boyle first published details of these experiments in *A Defence of the Doctrine, Touching the Spring and Weight of the Air*, an appendix to the second edition of *New Experiments ... its Effects*. It came during an extraordinarily intense period of experimentation and writing on his part, which included *Some Considerations touching the Usefulness of Experimental Natural Philosophy* in 1663, *Experiments and Considerations touching Colours* in 1664, *New Experiments and Observations touching Cold* in 1665 and two books in 1666, *Hydrostatical Paradoxes* and *The Origin of Forms and Qualities*. These books brought science to the middle classes and were lauded by the diarist Samuel Pepys for their clear, comprehensible prose. Boyle ensured they were published also in Latin for international scholars, while his religious romances, published at around the same time, were translated into several languages and became popular throughout Europe.

In the course of publishing these books, which were taken up and championed by the Royal Society, Boyle announced a string of notable firsts and discoveries, his methods becoming exemplary of the empirical method that the Society espoused. Like his vacuum pump, which after the Royal Society received its charter became a centrepiece of the Society's meetings, often being demonstrated in the presence of Charles II to foreign dignitaries and visiting monarchs, many of Boyle's groundbreaking experiments were first publicly demonstrated in front of the members of the Royal Society.

Boyle was the first chemist to collect a gas; he was the first to distinguish between acids, bases and neutral substances; he was the first to use coloured acid-base indicators; he showed that water expanded when it froze; he came very close to being the first person to discover a new element when he extracted phosphorus from urine but he failed to identify it; he pioneered and named the process of chemical analysis, thereby developing the ammonia test for copper and the silver nitrate test for salt in water.

Many of these discoveries were reported in the Royal Society's *Philosophical Transactions*, edited by Henry Oldenburg. It is perhaps unsurprising that Oldenburg devoted so much effort to publicising Boyle's work. In 1653, Oldenburg, a diplomat from the

Hanseatic city of Bremen in what is now northern Germany, was stranded in London when Sweden forcibly subsumed Bremen. When he heard of Oldenburg's predicament, Boyle employed him as a tutor to his sister's son, and later as an 'intelligencer' to manage his correspondence and dealings with foreign scientists. When the Royal Society was established, Boyle suggested Oldenburg as the Society's secretary for physical sciences.

In spite of his many ties with the Royal Society and the other early experimentalists who met there, Boyle remained in Oxford until 1668, when he moved in with his sister, Katherine, now Lady Ranelagh, in London. As in the 1640s, when Katherine had accommodated meetings of the Invisible College, the house in Pall Mall became a salon for London intellectuals, including the diarist John Aubrey, who described Boyle at this time as 'very tall (about six foot high) and straight, very temperate, virtuous and frugal, a bachelor, keeps a coach, stays with his sister. His greatest delight is chemistry.'

In 1680 Boyle declined the offer of the post of President of the Royal Society on religious grounds. Although he never saw any conflict between his religious belief and his mechanistic, philosophical and scientific views, religion was such an essential part of his life that he felt he could not swear the oaths necessary to serve as president.

Shy and sometimes awkward, Boyle never married, preferring to see out his days experimenting, accumulating information and writing in Pall Mall until he died a week after his sister in December 1691. By then, the Royal Society had progressed from a social club for gentlemen scientists to the leading scientific academy of the Western world, and the intellectual tussles of Boyle's former assistant, Robert Hooke, with Isaac Newton had made British science pre-eminent and raised the question of who of them was the greatest scientist of all.

Chapter 4

THE GREATEST SCIENTIST OF ALL

For more than 300 years one man ruled supreme in the pantheon of great British scientists: Isaac Newton. Even in the company of Charles Darwin, Stephen Hawking or Michael Faraday, Newton would be regarded as a cut above, possibly the greatest intellect the world has ever seen. And yet, any serious and unbiased study of Newton's achievements cannot avoid the uncomfortable fact that Newton's reputation relies in part on his deliberate obscuring and denial of the achievements of his biggest rival, a man whose contributions to science some historians regard as at least as significant as Newton's own. Newton's rewriting of history was so successful that it is only in the last few decades that Robert Hooke's reputation has been restored and the breadth of his achievements fully recognised.

So what is the evidence? Judging solely by their publications, Newton far outshone Hooke. Both Newton's *Principia* and Hooke's *Micrographia* were masterpieces. But without doubt, *Principia* was a far more significant work, justifiably regarded as the single most important book in the history of science.

With *Principia*, Newton not only unified mathematics and science; he also showed that the same physical forces apply throughout the universe, whether to a falling apple or an orbiting planet, from the microscopic to the cosmic. In explaining nature through a few relatively simple mathematical rules, Newton changed our perception of the universe from mystical to mechanistic. But most of all Newton was entirely correct. Unlike predecessors such as Descartes or Galileo, Newton got his mathematics and physics absolutely right. Even 200 years later, scientists were discovering new phenomena that were explained perfectly by Newtonian physics.

Although Hooke invented the field of microscopic biology with *Micrographia*, opening up a previously unseen realm to the gaze of the world, his book is not in the same league as *Principia*, which established fundamental truths about the universe. But the question that has dogged scientists and historians is whether Newton and Hooke should be judged only on the content of their books. If not, then what about their contributions to science through their other achievements and their relationships with other scientists? The answers are not straightforward.

Hooke was undoubtedly the greater experimentalist, his investigations ranging over a much wider expanse of scientific endeavours. Newton was the better theorist, but he drew (his detractors would say borrowed or even stole) much of his material from his rivals' work, not only from Hooke's experiments, but also from Edmond Halley, Christopher Wren and John Flamsteed. Newton's forte was his facility with mathematics, the most innovative elements of which he invented, such as calculus. This extraordinary talent enabled Newton to turn half-realised concepts, which his peers had been struggling to describe fully, into concrete universal rules that withstood experimental and theoretical examination.

That said, Newton might never have developed some of his theories if he hadn't been able to build on the work of others. In particular, Hooke was the first to propose that an inverse-square effect influenced the motion of planets, which Newton seized upon to develop his law of universal gravitation, his most significant single achievement. It's also beyond doubt that Hooke and others who were in correspondence with Newton suggested ideas that became cornerstones of Newtonism, and that Newton was often extremely reluctant to acknowledge the input of his rivals and peers.

Without doubt, Newton worked assiduously to denigrate Hooke's reputation. As a result, for two centuries Hooke was widely regarded as a deeply flawed and bitter character, and not the seventeenth century's single greatest experimental scientist that he undoubtedly was.

Hooke's first biographer portrayed him as 'in person, but despicable' and 'melancholy, mistrustful and jealous'. Subsequent writers said Hooke was 'cantankerous, envious, vengeful', 'suspicious' and 'irritable' with a 'cynical temperament', a 'caustic tongue' and 'uneasy apprehensive vanity'.

And yet, Hooke collaborated successfully with a range of scientists, including Boyle and Wren, maintained several deep and long-lasting friendships, was very supportive towards family members, and was a frequent attendee of social events. Meanwhile, Newton was an arrogant, some would say paranoid, loner who tolerated no disagreement or dissent.

* * *

Born the youngest of four children in 1635 to a modest family on the Isle of Wight, Robert Hooke was a weak and sickly child prone to debilitating headaches and, by his own admission, incapable of exercise more demanding than 'running and leaping'. Thinking he might not reach adulthood, Hooke's father, a curate, gave up on his son's education, leaving young Robert much to his own devices. However, from a young age Hooke was remarkably curious about the world and quick to learn. With a talent for making things work, he constructed his own toys, including a working wooden clock, which he designed himself after dismantling a brass timepiece, and a large model ship with fully operational guns and sails.

Hooke's father fell ill when Robert was 10, leaving his son to educate himself entirely. Freed from pursuing a formal education, Hooke let his imagination and talents run wild and was soon applying his exceptional observational skills to examining and copying the plants, animals, rock formations and sea scenes he'd seen around his home. He taught himself to paint and to draw by imitating the techniques of a visiting portrait painter, John Hoskyns, and made his own paints from coal, chalk, ruddle and clay. When Hooke's father died in 1648, his family decided he should use his £40 inheritance to move to London to take up an apprenticeship at the studio of the painter Sir Peter Lely.

Hooke's artistic training was short-lived. The paint fumes exacerbated his headaches and he decided he could teach himself more than Lely knew, so he bought himself an education at Westminster School, where in addition to learning Latin, Greek, Hebrew and oriental languages, he mastered all the mathematical teachings of Euclid in a single week. Recognising the exceptional talents of Hooke, the headmaster, Dr Richard Busby, encouraged him to pursue his own interests in much the same way as he had done before he arrived at the school. Given free rein yet again, Hooke combined his fascination with mechanics with his recently acquired understanding of geometry and was soon designing machines of his own, including a prototype flying machine. He also branched out into music, learning to play the organ.

By 1653, Hooke had decided that he'd learned everything that Westminster might teach him, so he took up a poor chorister's place at Christ Church, Oxford University. In those puritanical Parliamentarian days, the Anglican Church had been abolished and consequently choir attendance was suspended, so Hooke effectively received a small income simply for attending the college, which he supplemented by performing minor servant duties to a wealthier student, Mr Goodman.

Taught astronomy by Seth Ward and mathematics by John Wilkins, Hooke soon impressed his tutors with his natural ability, his skills and knowledge, and – working with Wilkins – his experiments with flying machines in the grounds of Wadham College.

Hooke's reputation appears to have preceded him from Westminster. Almost as soon as he arrived at Oxford, he was asked to join the Experimental Philosophy Club. Through it, he met Christopher Wren, who was about to receive his MA at Wadham. Although their backgrounds were very different – Wren came from distinguished Wiltshire landowning gentry and was three years older – their relationship prospered on a meeting of brilliant minds. Hooke had only just begun to explore the finer details of the 'new' experimental philosophy, but through his natural curiosity and mechanical genius he shared a passion with Wren for a range of interests stretching across anatomy, astronomy, chemistry, instrument design and architecture.

Hooke's studentship at Christ Church and his membership of the Experimental Philosophy Club brought him into contact with Thomas Willis, who by 1653 was in his Oxford heyday, working as a doctor, lecturing at Christ Church and conducting the dissections that led him to discover the functions of the brain. Willis asked Hooke to become his assistant. Hooke then developed the microscopes that enabled Wren to examine brain tissue for the illustrations in Willis's book, *Cerebri anatome*.

While Hooke's work with Willis taught him the principles of experimentalism, conversations with his tutor, Seth Ward, led him to ponder ways of better timekeeping in order to record astronomical observations more accurately. A major impetus for this was a solution to the longitude problem, at that time one of the most pressing problems for a maritime nation such as Britain.

Hooke would soon return to timekeeping, but before that, in 1655, he moved to Robert Boyle's employ on the recommendation of Willis and over the following five years helped Boyle achieve many of his discoveries in chemistry and physics, the most notable example being Hooke's construction and operation of Boyle's vacuum pump. During this time he also worked out a correct theory of combustion, invented or improved various meteorological instruments including the barometer, and made important astronomical observations.

While conducting experiments with a pendulum for Boyle in 1657, Hooke had a brainwave that went back to his earlier contemplations on timekeeping. He realised that the energy that made a pendulum swing with regularity was essentially the same as the energy needed to tension and release a steel spring. Hooke then worked out that a mechanical watch could be made to keep accurate time if its swinging balance was controlled by a fine hair spring (now called an isochronal balance spring). This 'circular pendulum' invention by Hooke gave the spring in a watch the natural evenness of a swinging pendulum and brought accuracy to watches that had previously lost up to ten minutes a day. Over the next three years, Hooke worked on improving his design until he was ready to file a patent drawn up by Boyle. But shortly before filing the patent, Hooke withdrew it in a fit of pique when he realised that anyone who improved on his design would be able to file a new patent and thereby receive royalties on an invention derived from his own discovery.

The patent, covering one of the first applications of a physics discovery in a technology product, would have made Hooke a very rich man. Instead, his reluctance to exploit or capitalise on his idea had the effect, some historians have suggested, of making him guard his future inventions and discoveries far more jealously, which in turn led to the disputes with Newton that damaged his reputation for centuries. As for the spring-controlled watch, Christian Huygens, a Dutch physicist, is usually credited with its invention in 1675, even though it was based on the same spring technology that Hooke first developed. The only legacy of Hooke's invention is the principle of elasticity (which says the extension of a spring is proportional to the applied force), now universally known as Hooke's Law and discovered by him in the course of developing the balance spring, but not publicised until 1678.

By the time that Hooke had decided not to patent his watch, the end of Britain's flirtation with republicanism was dawning and order was returning to London. Parliamentary troops stationed at Gresham College had left the capital, allowing members of the Oxford Experimental Philosophy Club to return to London to reinstate their meetings at the college, initially under the title of the Society for the Promoting of Physico-Mathematical Experimental Learning.

Hooke delivered a paper on capillary action at one of the Society's early meetings in 1661. The next year the Society received its royal charter, which contained a provision for appointing a Curator of Experiments to the newly established institution, now called the Royal Society for the Promoting of Physico-Mathematical Experimental Learning. Hooke was appointed to the post, initially unpaid, although he was later offered £30 a year, elected a Fellow of the Society and given the post indefinitely, making him effectively the first professional research scientist. From 1664, he was also appointed Cutlerian lecturer on mechanics and Professor of Geometry at Gresham College, posts that paid him £50 a year, although there were times when he struggled to get paid and had to rely on the generosity of Boyle to keep him afloat.

Hooke's task at the Royal Society was to demonstrate three or four major experiments at each week's meeting, a daunting assignment, but one to which he immediately rose, performing it excellently for forty-one years until his death.

In many ways, the post was ideal for Hooke. It suited him to apply his unparalleled mechanical and observational talents to hundreds of different experiments every year, but its disadvantage was that it gave Hooke little time to develop fully his most interesting ideas and findings. Almost all Hooke's later publications (and those of several of his peers and rivals) were rooted in the demonstrations he conducted in the early days of the Royal Society. They ranged from investigations into the nature of air, including the finding that it contracted when cooled and that it was essential to life, to experiments on pendulums, falling objects, gravity and weight. He devoted time to various aspects of the weather, including developing several meteorological instruments, such as the wheel barometer, rain gauges, hygroscopes for measuring humidity, a wind speed instrument and a weather-clock that automatically recorded various meteorological instrument readings. In some cases, such as his experiments with a 200ft-long pendulum in the steeple of Old St Paul's, it was physically impossible for Hooke to repeat the demonstration within the confines of the Royal Society, so instead he prepared a lecture to report on his method and findings.

In spite of his obligations to the Royal Society and Gresham College, Hooke found time to publish *Micrographia* in 1665. Described by Samuel Pepys, who stayed up into the early hours of the morning with it, as 'the most ingenious booke that I ever read in my life', *Micrographia* turned the power of optical magnification from the telescoped heavens to the realm of the microscopic. Just as Galileo's telescopic gazing into the sky had permanently changed humankind's perception of its place in the universe, so Hooke's journey into the microscopic opened a door on a parallel world that few people realised existed.

Micrographia shocked the world with its amazing sketches of tiny animals, minerals and plants. The first time anyone opened it, they must have been stunned by the world of the very small. Nobody had any idea about it before.

The exquisite drawings by Hooke showed amazing detail. In an illustration of a flea, we could see the hairs on the legs and the little hooks on the ends of their feet used to cling onto the host.

In another beautiful picture of a fly's head, the surfaces of the eye are arranged in beautiful geometric rows. Few had any idea that there was such precision, such geometric accuracy at the level of the very small. And Hooke was one of the first to see this.

This new minute world which so thrilled seventeenth-century society was only possible because of a new scientific instrument which Robert Hooke had perfected: the compound microscope with two lenses, the objective lens and the eyepiece lens.

I can imagine it must have been a wonderful experience for Hooke and people like him to look down a compound microscope for the first time, opening up a whole new world of the very small that nobody had ever dreamed of before.

For me as a biologist, what is remarkable about Hooke's drawings is not only their incredible detail, but how well they have stood the test of time. Now that we have scanning electron microscopes, we can look at the same things that Hooke studied, such as a fly's head, and see how accurate his pictures really are.

In its time, the world that Hooke unravelled for everyone was as wonderful as any view of the microscopic world we have now. And Hooke's own ingenuity in making and designing the instrument to see this was quite astonishing in the seventeenth century.

Richard Dawkins

Micrographia was an immediate bestseller. It brought Hooke scientific fame throughout Western Europe and had a profound influence on our understanding of the natural world. However, the full title (*Micrographia, or some physiological descriptions of minute bodies made by magnifying glasses, with observations and inquiries thereupon*) told only half the story. As well as describing hundreds of animate and inanimate objects, Hooke discussed many of the experiments he'd conducted at the Royal Society, adding details of his theories, evidence, results and conclusions.

Using his magnificent drawing and observational talents, Hooke had produced a series of illustrations that impressed with their accuracy as much as with their beauty. The images were supported by clear and lucid prose in which Hooke described hundreds of items, among them insects, sponges, bird feathers and butterfly wings, the compound eye of the fly, mosses, plankton, mica, crystals in flint, cork, kidney stones, fossils and charcoal. These observations led him to speculate and to devise experiments to test his speculations, which again he described in detail.

Observations of colour patterns in mica, for instance, led to a discussion of a theory of light, including how white light could be refracted through a prism into its constituent colours. Hooke also suggested that the colour patterns often seen in nature, such as on an insect wing or in oil floating on water, were caused by interference patterns in light, which implied that light was transmitted in waves.

While discussing his observations of petrified wood and cork, Hooke became the first person to use the word 'cell'. Describing small regular compartments in cork less than a thousandth of an inch across, he coined the term because of the compartments' resemblance to monks' cells in a monastery.

'I could exceedingly plainly perceive it to be all perforated and porous, much like a Honey-comb, but that the pores of it were not regular,' Hooke wrote. 'These pores, or cells, were indeed the first microscopical pores I ever saw, and perhaps, that were ever seen, for I had not met with any Writer or Person, that had made any mention of them before this.'

Writing on his observations of fossils, Hooke proposed a mechanism by which fossils formed (previously they were thought to be rocks that coincidentally resembled shells or small creatures) and suggested they could be used to look into the history of life on Earth.

In his examination of charcoal, Hooke suggested a mechanism for combustion that involved the combustible material mixing with air. He went on to discuss heat and cold, suggesting a temperature scale that set the freezing point of water at zero. *Micrographia* ends with a section on astronomical observations of stars and the moon.

If I reflect upon the importance and significance of Hooke on the history of biology, I realise there is a similarity to Darwin. Darwin wasn't the first to think of evolution but he was the first to realise the importance of evolution; to hit the world with the idea that evolution was the explanation for just about everything about life.

Hooke similarly wasn't the first person to invent the compound microscope, or the first person to make one, or to look down one.

But he was the first person to really show the world the importance of what you could see down the compound microscope. To open the windows to the world. To see into the miniature universe of things that no one had ever seen or dreamt of before.

I think there is a fascinating parallel between Hooke's work in the seventeenth century and what Darwin achieved in the nineteenth century.

Richard Dawkins

The extraordinary diversity of scientific investigation exhibited by Hooke in *Micrographia* hardly slowed after its publication. In May 1666, he presented a paper at the Royal Society that put forward the idea of an attractive force exerted by the Sun on its surrounding planets. Although Hooke did not call this force gravity, he suggested that it somehow held the planets in their orbits, which was a stark departure from the orthodox view put forward by Descartes that whirlpools in the ether kept the planets spinning around the Sun.

Hooke did not return to his considerations of gravity for another eight years, mainly because he had even less time available after September 1666, when his great friend Christopher Wren appointed him an official surveyor in the rebuilding of London following the Great Fire.

In 1666, London burned like rotting sticks. For four days and three nights, flames engulfed the city. While Hooke could have watched as the flames stopped only a street away from the Royal Society, in Oxford Wren saw the sky glowing red in the east and recognised a god-given opportunity.

Like most scientists of his time, Wren was a polymath. He was interested in everything from the movements of the Moon to the behaviour of bees. He had an insatiable curiosity and he used it to investigate everything he saw around him.

Wren was inspecting the ruins while the embers were still hot. More than 12,000 homes were burnt to the ground and three-quarters of the city was destroyed. Wren saw his chance and within days he presented personally to the king his plans for a new city.

Wren's great vision was never fully accomplished, but nevertheless, with Hooke as his right-hand man, he created and designed some of the most ambitious and innovative buildings that London had ever seen. Buildings fit for a new scientific age.

David Attenborough

For Hooke, the new position was very welcome, not least because it paid him a decent wage, but it meant he no longer had the time to research to the same extent that had enabled him to produce a publication of the magnitude and breadth of *Micrographia*. Nevertheless, he continued his relentless schedule of three or four experiments or demonstrations at the Royal Society each week, while simultaneously taking on one of the most demanding jobs in London. Working as Wren's assistant, Hooke proposed a new gridded street layout within a week of the fire. He re-established property lines, supervised rebuilding and designed several buildings, including the Monument, which he designed with an open interior column so that he could conduct experiments on falling objects. Other designs included Montague House, the Royal College of Physicians and Bethlem Hospital (the psychiatric sanatorium later nicknamed Bedlam) at Moorfields. He also assisted Wren in the design of the Royal Greenwich Observatory and St Paul's Cathedral, applying his mechanical skills to devising a method of construction for its astonishing unsupported cupola dome 300ft over the ground and totally unlike anything seen before in England. It was called a triumph of converting pure geometry into mechanics.

Meanwhile, Hooke exploited his position as surveyor to design a remarkable telescope within one of the buildings he was designing for the new London.

In many ways, the astonishing burst of scientific growth that created the modern age started with an obsession with measuring, and Robert Hooke made one of the most powerful and largest scientific instruments ever made.

His 200ft parallax telescope is still right at the heart of the City of London. It's the Monument: the tallest isolated stone column in the world, built in 1671 to commemorate the horrors of the Great Fire.

But Hooke designed it so that he could use its great height to peer into the night sky, measure what he saw, and begin finally to understand the behaviour of the stars, planets and mysterious comets.

At the base of the tower was the eyepiece of Hooke's telescope. Hooke actually lived in the basement while the tower was being built. And that's where he would have lain underneath the eyepiece, which was a little lens. At the far end, right at the top of the tower in the flaming urn, which had a hinged lid, would have been the objective lens. The two lenses together, the objective lens at the top and the eyepiece, constituted the longest, most powerful telescope that had ever been built or conceived until then. Although it never worked – the structure swayed in the wind and expanded and contracted as the temperature changed – it marked a seismic shift in scientific ambition.

Richard Dawkins

In 1672, Hooke's niece, Grace, came to live with him in London. Ostensibly she was joining him as a housekeeper, but as Hooke's detractors are fond of pointing out, Hooke had a history of sexual relationships with his maidservants, and eventually, at the age of 15 or 16, Grace followed them in becoming his mistress. Coincidentally, 1672 was also the year the confrontation between Hooke and Newton started.

Secretive, introverted and more than slightly paranoid, Newton had by 1672 been working on his own for almost a decade at Cambridge, which at that time was an academic backwater with very little of Oxford or London's reputation for scientific endeavour. Newton, in spite of keeping his investigations of light, colour, calculus and gravity very close to his chest, had sufficiently impressed the university authorities to be appointed a Fellow in 1668 and Lucasian Professor of Mathematics in 1669. Two years later, details of Newton's work on splitting light through a prism and his design of a reflecting telescope reached London and Newton was invited to the Royal Society. Immediately after the telescope was demonstrated at Gresham College, Newton was elected a Fellow of the Royal Society and asked to submit other examples of his work. Early in 1672, he presented a paper on light and colour that included details of his splitting of white light through a prism into its constituent colours.

Recognising that Newton's work on light was based on some of the ideas he had put forward in *Micrographia*, Hooke was outraged. Although Newton's experimental method had produced a clearer spectrum with a more pronounced spread of colours, Hooke was irritated that he had not been acknowledged or credited by the young upstart, who had made little more than a passing reference to Hooke's earlier work on interference patterns in light. Adding insult to accusations of plagiarism, Hooke said that the parts of

Newton's theory that had not been stolen from him were in fact wrong. Given that Newton believed that light was made up of tiny particles, not waves, we now know that Hooke's accusation was correct, but at the time Newton's incorrect corpuscular theory of light appeared to be as valid an explanation as Hooke's correct wave theory. Now, Newton was just as outraged as Hooke.

Relations between Newton and Hooke soon took a further turn for the worse when Oldenburg, the joint Secretary of the Royal Society, passed on an exaggerated account of Hooke's outrage at Newton's appropriation of his work. For Newton, who even as a young man had an extremely high opinion of his own achievements and little or no respect for the accomplishments of others, any criticism was very unwelcome. The feud festered on until 1673, when Newton threatened to leave the Royal Society and to cease open publication of his work. Only a fawning performance by Oldenburg and an offer to waive dues to the Society persuaded Newton not to leave.

While the feud with Newton quietly simmered in the background, Hooke did his reputation for awkwardness no favours two years later when, in 1675, Huygens revealed his invention of the balance-spring watch, which – as mentioned earlier – used exactly the same technological principle as the timepiece that Hooke had failed to patent eighteen years earlier. Hooke fired off an invective-filled letter, with accusations that his ideas had been stolen. It achieved very little. Huygens was the first to publicise the discovery and consequently Huygens has been credited with its invention ever since. Hooke's only recourse was to claim belatedly (and somewhat petulantly) that the balance-spring was his invention in a postscript to his publication, *Description of Helioscopes*, later that year.

By 1676, Hooke's dispute with Newton had been running for four years and the Royal Society was starting to look more like a forum for petty squabbles than a centre of scientific excellence. That year, Thomas Shadwell published *The Virtuoso*, satirising the Royal Society and its members. Recognising the character of Sir Nicholas Gimcrack as a parody of himself, Hooke wrote in his diary after attending a performance: 'Damned Dogs. *Vindica me Deus* (Grant me revenge, God). People almost pointed.' Clearly the time had come for the Royal Society to act if it was going to protect its reputation. The Fellows forced Hooke and Newton to enact a public reconciliation in the form of an exchange of letters.

Hooke took a conciliatory line in his letter, magnanimously acknowledging Newton's achievements, gently pointing out his own prior work, then seeking resolution, writing that 'your design and mine, are, I suppose, both at the same thing, which is the discovery of truth'.

In his response, Newton wrote the phrase for which he has become famous and over which historians of science have debated ever since: 'If I have seen farther, it is by standing on the shoulders of Giants.'

This appeared to be an acknowledgement at last of Newton's debt to Hooke. But knowing now that Newton frequently bore grudges, and with the weight of evidence showing that he subsequently did everything in his power to damage Hooke's reputation, this interpretation seems less likely, particularly as Newton prefaced his acknowledgement with a reference that suggested Descartes had an earlier claim to priority than Hooke. No one but Newton knew his exact intention, but Hooke's supporters point at the capital G used by Newton for 'Giants' and suggest that it was an intentional dig at Hooke's small, hunched stature, implying that Newton had no need to build on the work of a small (mentally as much as physically) man like Hooke.

Although everything was settled in public (Hooke and Newton exchanged cordial letters on the refraction of light the next year), in private the damage had been done. Abiding by his threat to withdraw, Newton was not heard of publicly for eleven years until 1687, when he published his three-volume masterpiece, *Philosophiæ Naturalis Principia Mathematica*. As for Hooke, on the death of Oldenburg in 1677, he succeeded to one of the two posts of Secretary of the Royal Society, which he held while still maintaining his responsibilities as Curator. However, he would clash with Newton again several years later.

In the meantime, in 1678 Hooke published his treatise on elasticity, *De potentia bestitutiva* (*Of Spring*). It contained the first public enunciation of Hooke's Law, which he'd discovered while working on his spring balance watch some twenty years earlier. The next year, Hooke wrote to Newton, asking him to comment on some thoughts he'd had about gravity which originated in a lecture he'd first given in 1674. In that lecture, he'd put forward three suppositions, the first of which suggested that all bodies have an attractive force 'whereby they attract not only their own parts … but that do also attract all other Coelestial Bodies that are within the sphere of their activity'.

Although Hooke hadn't used the word 'gravity', he was stating its existence, a remarkable leap of imagination. At a time when scientists were trying to explain the motion of planets in terms of a centrifugal (that is, *escaping* from the centre) force, Hooke had inverted the popular theory of circular motion by focusing attention on a centripetal (that is, centre *seeking*) force that continually deflected moving planets into their orbits.

Next, Hooke put forward a theory in his second supposition that would later be reiterated in almost the same form as Newton's first law of motion. Hooke told the audience at his lecture that 'all bodies whatsoever that are put into a direct and simple motion, will continue to move forward in a straight line, till they are by some other effectual powers deflected and bent into a Motion, describing a Circle, Ellipsis, or some other more compounded Curve Line'. Hooke's third supposition combined the attractive force suggested in his first supposition with the motion described in his second supposition, stating that 'these attractive powers are so much the more powerful in operating, by how much the nearer the body wrought upon is to their own Centres'. This suggested that the attractive force between two objects was inversely proportional to the distance between them. The only detail separating this statement from the principles expressed in the law of universal gravitation, derived by Newton several years later, was Hooke's failure to realise that the force was inversely proportional to the *square* of the distance. However in 1679, when Hooke wrote to Newton, he corrected that error.

After exchanging several letters, in which Hooke and Newton each commented on and corrected assumptions they felt the other had made erroneously, Hooke wrote that he thought that gravitational attraction acted 'in a duplicate proportion to the Distance from the Center Reciprocall'. There it was: the inverse-square law stated in ink on paper by Hooke to Newton, eight years before Newton published it in *Principia* as his Universal Law of Gravitation.

It's therefore not surprising that Hooke was outraged that Newton failed to give him credit for introducing him to the inverse-square law. Unfortunately for Hooke, he was unaware that Newton had not only postulated it independently and earlier than he, but had tested it by calculation in two different ways. Newton had simply not publicised his discovery, so when Hooke wrote to Newton insisting that he deserved some credit, Newton refused to reply and instead deleted any mention of Hooke from the third volume of *Principia*, which was about to be published. Hooke never got over it.

Hooke's health was already in decline after the death of his niece, housekeeper and lover, Grace, in 1684. Following the dispute with Newton, Hooke became increasingly bitter, cynical and solitary. He continued to conduct weekly experiments and demonstrations at the Royal Society, but produced no more groundbreaking discoveries, and frequently made vitriolic asides about Newton during his lectures. Blind and bedridden for the last year of his life, he died in 1703.

Given his contribution to science, his rare combination of talents and skills, mostly self-taught, and the way in which he turned the Royal Society into the world's most respected academy of science, Hooke would be quite justified if he felt aggrieved at not receiving adequate recognition for his achievements. His sense of injustice during his lifetime that others had stolen ideas that he had been first to put forward, or that they had built on his ideas without giving him adequate recognition, is entirely understandable. Few other scientists of the period produced as many original ideas over such a wide diversity of scientific subjects. But if Hooke had a fault it was his frequent failure to develop his ideas into comprehensive theories or to publicise them adequately. In part, this was due to lack of time. In some cases, such as the spring balance watch, it was because of a personal misjudgement. But just as often – possibly because he was largely self-taught – it was because he did not have the full technical or mathematical ability to turn his ideas into physical laws of nature.

As for the inverse-square law at the heart of Hooke's dispute with Newton: in fairness to Newton, Hooke had only suggested it, whereas Newton was the first to prove it mathematically.

Hooke had also mentioned it to Wren and Edmond Halley after a meeting of the Royal Society in 1684. They all knew Kepler's laws of motion and Huygens' 1673 treatise about the forces on an orbiting object that suggested an inverse-square relationship, but in spite of Wren issuing a forty-shilling wager to Hooke, neither he nor Halley had the mathematics skills to prove it.

That said, Hooke certainly didn't deserve Newton's petty and vindictive treatment, particularly as the historical evidence suggests that Newton might never have attempted to prove and establish the law had Hooke not written to him in 1679. At that time, Newton had abandoned all natural philosophising and was interested only in theology and alchemy. Without Hooke's prompting (and Halley's encouragement, as we shall see later), it is entirely possible that Newton might never have returned to natural philosophy, completed his greatest work, published *Principia* and achieved his reputation as the greatest scientist of all. Given those factors, it is perhaps unsurprising that Newton did everything in his power to obscure Hooke's achievements and tarnish his reputation for centuries.

Was Hooke's paranoia justified, or was his unrestrained ego wildly out of control? We can only say this: in the end, his career was blighted by an enormous fight over scientific priority that was fought over most of his adult life. And the reason most of us have never heard of him is that the fight was with someone rather better known: Isaac Newton.

Regarded by many physicists as the greatest scientist of all time, Newton had a dark flipside to his genius: a most unpleasant character. He was humourless and bad-tempered; he hated almost everyone and had few friends. It's even been said somewhat waspishly that he had to teach himself to smile by practising in front of a mirror.

* * *

Seven years younger than Hooke, Isaac Newton was born on Christmas Day 1642 to an upwardly mobile Lincolnshire farming family. Small enough to fit in a quart pot at birth, Newton's survival was unsure at first, but he eventually prospered.

His mother, Hannah, widowed six months before Isaac's birth, was devoted to her son for his early years. But in 1645, Hannah remarried and, in return for a piece of land bequeathed by her new husband to her son, she sent young Isaac to live with her parents.

Living with grandparents was a mixed blessing for Newton. No love was lost between Isaac, who never had anything good to say about his grandparents, and his grandfather, who pointedly omitted Newton from his will. Historians often attribute the worst characteristics of Newton's adult personality – his anti-sociability, paranoia, arrogance and general unpleasantness – to this formative period in which his emotional needs were neglected in the absence of maternal love. However, his grandparents' eagerness to send Isaac to school to get him off their hands meant he gained an education that, as a son of an illiterate farming family, he would otherwise have been denied.

After a primary education, Newton was sent to a grammar school in Grantham when he was 12, lodging with the Clark family. Mr Clark was an apothecary; his wife had a brother, Humphrey Babington, who was a Fellow at Trinity College in Cambridge and who would later become a significant influence in Newton's academic and research career.

Newton did well at school. Like Hooke, he was somewhat solitary and skilled at building models, on one occasion frightening the local populace by flying a glowing lantern from a kite in the night sky. With the supportive encouragement of his head teacher, John Stokes, he soon mastered the classics, then started taking an interest in mathematics. But when Newton was 16 his mother removed him from school. She wanted young Isaac to be taught how to run the family farm. Whether by intent or not, Newton made a complete hash of his agricultural training. His incompetence was so extreme it became local legend. Assigned tasks such as shepherding a flock, he buried his nose instead in textbooks or built waterwheel models, absentmindedly allowing the sheep to escape and graze on a neighbouring farmer's crops.

Eventually Newton's ineptitude persuaded his mother to listen to various voices suggesting that Newton should return to school to prepare for university, among them her Cambridge graduate brother, Babington, and Stokes, who was so impressed with Newton's potential that he offered to waive the fees if the boy came back to school.

In 1661, at 18, a late age compared to his peers, Newton arrived at Trinity College, Cambridge. Newton was heir to the lordship of a manor and son of a wealthy widow, but his mother, resentful of an education she regarded as a waste of money, allowed him only £10 of her £700 annual income, so Newton paid for his keep by signing on as a sub-sizar. It meant performing menial tasks, such as cleaning and emptying out chamber pots, and dining separately from fee-paying students, a stigma that may have played a role in Newton's later social isolation at the university.

Whatever the ramifications of his social standing, Newton was doing well academically, even if it was by following a typically unconventional path. After a short period reading the classics of natural philosophy, primarily Aristotle, he abandoned his prescribed curriculum to study independently the works about the new experimental philosophy by Descartes, Galileo, Boyle and Hobbes. With no equivalent to the Experimental Philosophy Club, Cambridge had little of the intellectual vigour of Oxford, so again Newton was left to work alone. With no tutorial guidance whatsoever, Newton wrote a title on a page of his notebook – *Questiones quaedam philosophicae* (Certain philosophical questions) – and beneath it, in his spidery handwriting, he listed forty-five subject areas that he thought worthy of his investigation including colour, gravity, the motion of bodies, the expansion of gases, capillary action and surface action. How did the Sun, the Moon and, of course, comets actually work? Why were some things heavy and some things light? What was air? What was in it? How did it move? What happened when air was put under pressure or removed? Like Hooke, Newton realised he wanted to know what light actually was.

The rigorous and obsessive experimental techniques of Robert Boyle, and the groundbreaking optical work of Robert Hooke, set the agenda for Newton's own work. Virtually unknown, dismissive of others, and grindingly poor, Newton set his scientific sights ambitiously high. He embarked on a series of stomach-churning experiments which would both announce his arrival in the scientific world and bring about a catastrophic rivalry that would last thirty years and destroy Robert Hooke.

A remarkable period of groundbreaking work followed. Obsessive at the best of times, Newton would often forget to sleep or eat when he was studying or undertaking experiments. But with the prospect of graduation approaching in 1665, Newton needed to find a way of staying at the university to pursue his vocation.

With the help of a recommendation from Babington and in spite of his unconventional studentship, Newton was elected to a scholarship, the first step to a fellowship. It provided free accommodation, a small income and liberation from the ignominy and drudgery of being a sub-sizar. It also allowed him to remain at the university for three years after graduation. With his future guaranteed, Newton embarked on a period of discovery subsequently called his *annus mirabilis* (wonderful year), although given that technically it started in mid-1665 and ended some time towards the end of 1666 or possibly in early 1667, *anni mirabiles* would be a more appropriate description.

In the summer of 1665, shortly after Newton received his BA, plague hit Cambridge and the university dispersed. Newton returned to his Lincolnshire home, Woolsthorpe, until April 1667. This long period at home – except for three months in 1666 when Cambridge was thought to be free of plague – gave Newton the seclusion needed to explore the capabilities of his own mind. Although he ranged across a wide range of science, at the core of his exploration was an unprecedented leap in Western mathematics. Entirely self-taught, Newton had spent a year at Cambridge forcing himself to ingest the teachings of Descartes, Oughtred and Wallis, among several others. Now at home, he was ready to make discoveries of his own, which he detailed in a 1,000-page notebook that he called his waste book.

The first breakthroughs were his discoveries of the laws of integration and differentiation, which led to his invention of the fundamental theorem of calculus – although in typical Newtonian fashion he would clash several years later with the German mathematician Gottfried Leibniz over priority. Newton was first, but kept his discovery to himself; when Leibniz published in 1684, Newton accused the German of plagiarism. In spite of only cursory evidence – Leibniz made a short visit to London in 1676, when he *might* have been shown Newton's unpublished work on calculus – the Royal Society launched a formal investigation in the eighteenth century. Although it upheld Newton's claim, evidence from Leibniz's notebooks shows the German to have started work on calculus before his visit to London and to have developed it independently from Newton. The dispute created a long-lasting rift between British and continental European mathematicians that held back British mathematics for more than a century.

After inventing calculus, Newton made several other discoveries in mathematics, which he summarised in a paper in November 1665. In a few short months, he had achieved more than enough to become the world's paramount mathematician (although, having not published his work, few people realised the extent of his achievements at the time). He updated the paper six months later and again in the autumn of 1666, but other than that he considered his work on maths done and he moved on to other scientific investigations.

Of Newton's many other discoveries, it is impossible to say which was the most significant, simply because so many were extraordinary, but probably the best known is the one that involves the story of a falling apple.

According to Newton's account related in old age to William Stukeley, one of his first biographers, he was sitting in his garden at Woolsthorpe in the summer of 1666 when he saw an apple drop from the tree. Realising that the cause of its fall was Earth's gravity and that the pull of gravity extended to any apple on any branch of that tree, Newton wondered whether Earth's gravity might also extend far enough to exert a pull on the Moon. If that was the case, thought Newton, then the power of this universal force to attract both a nearby relatively light apple and the distant but vast mass of the Moon might be explained by an inverse-square relationship.

Whether Newton's realisation of the inverse-square relationship of gravitation occurred exactly as described to Stukeley has intrigued scientists, historians and philosophers for centuries. If true, then Newton had devised the inverse-square law long before Hooke, Wren or Halley discussed it. However, there is evidence neither of the apple incident nor of thoughts concerning gravity or the Moon in Newton's contemporaneous notes – and we now know that Newton was no stranger to self-mythologisation.

The first mention of the Moon in Newton's notes appeared in the late 1660s, when having worked out a possible measure of gravity at the Earth's surface Newton mused about a possible link, through Kepler's third law of planetary motion, which suggested an inverse-square relationship between the maximum distance of an orbiting planet and the duration of its orbit. However, at this point there was no proof at all that Newton had formulated his law of universal gravitation.

However, Newton made other remarkable discoveries in the period of his plague exile from Cambridge, such as his work on mechanics, in which he looked at the effects and behaviour of colliding bodies, which would eventually lead him to his laws of motion.

Writing in his notebook, he calmly listed his achievements during this period, any one of which would have been enough to establish his reputation as one of the greatest scientists ever, but as a complete body of work they put Newton in a class of his own.

'In the beginning of the year 1665 I found the method for approximating series and binomial theorem,' Newton wrote. 'The same year I found the method for tangents of Gregory and in November had the direct method of fluxions [Newton's term for differential calculus] and in January had the theory of colours and in May following I had entrance into the inverse method of fluxions [integral calculus] and in the same year I began to think of gravity extending to the orb of the moon.'

In the field of optics, his most notable work was his splitting of white light through a prism into a spectrum of its constituent colours, which he then recombined into a beam of white light. Newton performed essentially the same experiment as described by Hooke in *Micrographia*, but by recombining the spectrum he proved the colours were not a property of the prism, but were contained in white light itself. And where Hooke had produced a white beam with blurred coloured edges, Newton produced a clear spectrum with defined edges between the colours. He achieved this by projecting the beam onto a wall 22ft away, which gave the various colours a greater distance in which to disperse. Given that, at the earliest, Newton did not start his work on optics until late 1665, the year in which Hooke published *Micrographia*, Hooke's complaint that his work was not credited appears justified, particularly as Newton abandoned further work on optics after his experiments with splitting white light.

However, Newton's work with spectra led him to realise that the refracting telescopes in use at the time would never be able to produce clear images of greater magnification, simply because their lenses and prisms had a tendency to split light into spectra, a characteristic called chromatic aberration.

Newton's response to the limitations of refracting telescopes came after he returned to Cambridge in April 1667, when he was elected a minor Fellow of Trinity College. He was obliged to teach more, but being almost pathologically shy he found lecturing students a painful and horrifying experience. So he made his lectures impenetrable, esoteric and downright boring, a tactic some academics are accused of employing to this day. His weekly classes quickly became empty, leaving him more time for what he really wanted to do: to lock himself away in his makeshift laboratory and study his new scientific questions – alone. In 1668, Newton invented a reflecting telescope that used a parabolic mirror instead of a lens to focus light, and although this was essentially a toy, six inches long and one inch in diameter, it magnified by up to forty times. It produced clearer, brighter images of much greater magnification than refracting telescopes, and the same essential design is used in the largest modern optical telescopes.

Although he had invented a much better telescope than any used previously, for Newton it was little more than a distraction from his pursuit of theoretical study backed by experimental proof. Ensconced in his rooms, he continued to study and write obsessively, noting one day that out of too much study and passion 'cometh madnesse'. Judging by some of the experiments that Newton performed on himself during this period, it would not be unreasonable to conclude that Newton's work was driving him close to insanity. On one occasion, he stared at the Sun for so long that he nearly went blind and had to spend several days in a dark room until his eyesight recovered. He also jabbed a bodkin – a fat needle used for stitching leather – 'betwixt my eye and the bone as near to the backside of my eye as I could' to investigate its effect on his colour vision. And on getting up in the morning, Newton would often sit for hours on the edge of his bed as if frozen, his attention so distracted by the thoughts in his head that he'd forget to stand up. The only external concern that impinged upon Newton at this time was a very real worry that his stay at Cambridge might end before he had completed his work. Fortunately he managed to avert that fate in 1668, when he was elected a major Fellow of Trinity College and granted his Master's degree. It earned him an income, free accommodation and at least another seven years' independent study at the university.

Around this time, Newton became more closely acquainted with Isaac Barrow, the first person to hold the Lucasian chair of mathematics at Cambridge. Barrow passed on to Newton a copy of *Logarithmotechnia* by Nikolaus Mercator, a mathematician who – having taught in Holland, Copenhagen and Paris – was now a member of the Royal Society, teaching in London. Stung by the fact that *Logarithmotechnia* contained details of mathematical functions that he had discovered several years earlier, Newton responded by writing *De analysi per aequationes numero terminorum infinitas* (*On Analysis by Infinite Series*), a tract that was passed on to John Collins, a mathematician in London who corresponded with many of the leading European mathematicians of the day. *De analysi* provided the first sign to the outside world of Newton's prodigious talents. Coming just a few months after Newton had described his reflecting telescope to Henry Oldenburg, it added to the growing clamour surrounding Newton, which culminated in his appointment as Barrow's successor to the chair of Lucasian Professor of Mathematics.

Despite the grandeur of the title, Newton appears to have been unimpressed by his professorship. Required by the appointment to teach, he gave the bare minimum of a lecture a week for one term a year. Often he spoke to an empty hall, but during some of the lectures he revealed the work on optics and mechanics that had consumed him for the past five years. When news of the lectures reached London, he was asked to send his reflecting telescope to the Royal Society, which in early 1672 elected him a Fellow. Soon thereafter, Newton published a letter on light and colours in the *Philosophical Transactions*. He followed it with an account of his reflecting telescope and seven additional papers on optics that year.

Having spent such a long period working in seclusion, the sudden welling of interest and attention surrounding his achievements would have been warmly welcomed by most people. But not Newton, who had already moved on from optics and mathematics, having become interested in chemistry in 1669. He was now consumed with alchemy and was also developing an obsession with theology. More significantly, like many successes suddenly thrust upon the public stage, Newton was finding himself surrounded by detractors, critics and accusers.

As well as Hooke's accusation that Newton's achievements in optics were indebted to his work published in *Micrographia*, Newton had to face criticisms of his theories from Huygens, regarded then as Europe's leading scientist. Other criticisms came from Franciscus Linus in Liège, the British-born monk who in 1660 had panned Boyle's barometer findings.

To fend off Linus, Newton offered the Royal Society two further papers on optics, which again drew waspish comments from Hooke, who claimed that much of the science had already been covered in *Micrographia*. Although Hooke and Newton soon patched over their disagreement (albeit only temporarily until they clashed again over the inverse-square law in 1687), Newton was sufficiently vexed by Linus' criticisms – which by 1675 had been taken up by one of Linus' students – to retreat altogether from public life.

'I see I have made myself a slave to Philosophy,' Newton raged in a letter to Oldenbury at the Royal Society, 'but if I get free of Mr Linus' business I will resolutely bid adew to it eternally, excepting what I do for my privat satisfaction or leave to come out after me.'

Newton lived up to his threat. By 1676 he had withdrawn to concentrate on alchemy and theology. For all intents, his days of scientific investigation were over. All his later scientific work – indeed most of his work after about 1669 – was revision and extension of his extraordinary creativity in the five years from 1664. Newton would not be heard from again for another eleven years, when he would interrupt his retirement from natural philosophising to publish his masterwork of the first scientific revolution, *Principia*, after which he stopped being a scientist altogether. And had it not been for the third great instigator of the final act of the scientific revolution – Edmond Halley – Newton might never have been cajoled to break his self-imposed purdah at all.

Chapter 5

A BRIGHT LIGHT IN THE SKY

Edmond Halley was the antithesis of Hooke and Newton. He was as gregarious as Newton was solitary, as handsome as Hooke was ugly. He drank brandy, smoked and swore like a sea captain. He had a passion for adventure, courted women with silks and satins, and clashed with pirates hunting treasure on the high seas.

The son of a wealthy businessman, Halley arrived at Oxford in 1673 after an education at St Paul's School in London during which he'd been blessed with all the toys an astronomy-obsessed child of an indulgent father could want. His father, also called Edmond, was a soap boiler at a time when the monopoly in soap production was controlled by the crown and heavily taxed in the form of the soap duty. Highly regarded in the City, Edmond senior was so wealthy he could afford to send his son off to university with a telescope and sextant better than those used by many professional astronomers.

At the time Halley arrived in Oxford, astronomy was regarded as much as a practical technology as a science of discovery. Before the advent of accurate clocks, it was impossible to determine a ship's position at sea accurately, particularly its location east or west – what's known as longitude. It meant vessels got lost, or worse, wrecked on rocks and reefs. Figuring out one's bearings at sea became the most burning economic, scientific and political issue of the time. And for seafaring nations searching for a reliable method of determining longitude, obtaining an accurate map of the stars was a matter of national importance.

In 1674, news reached London that the French had solved the latitude problem by using the relative positions of the Moon and stars. Concerned that Britain was being left behind, Charles II appointed John Flamsteed, a largely self-taught astronomer, to the post of the King's Astronomical Observator (the first Astronomer Royal) with the task of applying 'himself with the most exact care and diligence to the rectifying of the tables of the motions of the heavens, and the places of the fixed stars, so as to find out the so much desired longitude of places for the perfecting of the art of navigation'.

Meanwhile at Oxford, Halley had written a commentary on Kepler's laws of planetary motion, which he sent to Flamsteed in 1675. Impressed by Halley's work, Flamsteed invited the young lad that summer to the new Royal Observatory being built by Wren and Hooke on Greenwich Hill. As Flamsteed's protégé, Halley soon gained attention for his observations of two lunar eclipses and a sunspot, and for papers on planetary orbits.

At that time, astronomers believed the key to solving the longitude problem would be a more accurate map of the sky. When Flamsteed set about surveying the northern sky, Halley spotted an opportunity and put himself forward to map the southern sky. With his father's financial backing and the support of the king, Halley set off in 1676 aboard an East India Company ship for St Helena in the South Atlantic, at that time the most southerly British possession.

Halley's intention was to use a simple telescope and sextant to determine the positions of all the stars in the southern sky, a sky that had never been mapped before. Then, using his new map, the navy would be able to make the first decent stab at navigation in these treacherous waters. Mapping the sky would help Britain map the seas. And if Halley completed the task, it would make him rich and secure his reputation as an astronomer.

The voyage took many months across some of the roughest seas of the South Atlantic. And once Halley landed on shore, his troubles were far from over.

Halley found himself on a remote rocky island in a state of almost continual mutiny against a governor, Gregory Field, attempting to impose his sadistic will onto his population with an iron fist. St Helena had only three years previously been recaptured from the Dutch after they forcibly occupied the island and ejected one of Field's predecessors, Governor Anthony Beale, following Charles II's declaration of war on Holland in 1672. Administered by the English East India Company, the island was of enormous strategic and commercial importance. In 1659, it was seized under royal charter in order to ensure complete control of the trade route through the eastern South Atlantic. However, the governor was in constant conflict with the directors of the East India Company in London, who were more concerned with issues of profit and loss than the man at the sharp end, who ruled over a small population of company employees, English settlers and Goan, Malayan and Madagascan slaves.

Into this maelstrom stepped Halley. Fortunately, Charles II had ordered the governor to give him any assistance he needed and he promptly retreated to his makeshift observatory, out deep in the hills. Life was basic, the solitude mind-numbing and the skies often too overcast for astronomy. But over the course of the year, using his telescope and sextant, he accurately plotted the positions of no less than 341 stars, performing all the calculations and measurements himself. It was a heroic job. He returned to London two years later with the first catalogue of the stars compiled with a telescope. Although the map contained only a relatively small number of the southern sky's stars, Halley was deemed a great success. Regarded as a hero, he was elected to the Royal Society, honoured with an Oxford MA (he'd left without graduating), and declared the 'southern Tycho' in acknowledgment of Tycho Brahe, the last and greatest of the naked-eye astronomers.

His name made at the tender age of 22, Halley saw no great need to trouble himself immediately with further laborious observations of the night sky. Instead he interspersed travel to Danzig, France and Italy with socialising with Boyle, Wren and Hooke in London coffee houses, dining with visiting luminaries such as Peter the Great of Russia, nurturing his growing reputation as a ladies' man and attending meetings at the Royal Society.

While visiting Paris in the winter of 1680, Halley observed a comet that was visible to the naked eye in the night skies over much of Europe. Intrigued by the comet, he discussed it at length with various scientists and astronomers that he met during his European tour, including the great Italian-French astronomer Giovanni Domenico Cassini in Paris. On returning to England the next year to marry, Halley continued his studies of the night sky, hoping to crack a method of finding longitude from the position of the Moon. Working in the Octagon Room, designed by Wren with tall windows facing north from one façade to give a clear and wide-ranging view of the sky, an elaborate plaster ceiling and 20ft-high walls to fit elaborate pendulum clocks at the heart of the Royal Observatory in Greenwich, Halley observed the skies with the latest high-powered, more accurate telescopes and discovered something totally unexpected. Contrary to all the scientific opinions of the day, Halley found that Earth's constant companion, the Moon, was misbehaving. Instead of moving around the sky at a constant pace, it was speeding up. Not by much, but the new equipment was sensitive enough to notice. The lunar puzzle haunted Halley. He sat down and checked his results again and again, but he always came up with the same result. Not one of his theories could explain why.

But then something startling happened. The second comet visible to the naked eye in as many years appeared in the night sky over London. Halley spent many hours observing it from the house he and his wife had bought in Islington. Like many astronomers, he was searching for an explanation for the comet's sudden arrival, slow progression across the sky, and subsequent disappearance.

That autumn of 1682, as the comet lit the night sky, Hooke, Boyle and Halley would meet in coffee houses in London's fashionable Cornhill area to suggest new theories to explain the comet's behaviour.

As they looked up into the sky at the new comet, their scientific imaginations were tested. This comet was behaving even more strangely than the last. No sooner had it arrived than it disappeared, and then, a month later, it reappeared. It was unlike anything they had ever encountered before.

It was such a shocking thing for them. They were used to looking at the planets and the Sun and the stars and the Moon, trying to work out the laws of the universe, and suddenly this comet appeared from nowhere.

It made them focus on why these bodies might be moving in the way they did. They started talking about the forces involved, but at this time people believed that you could only move something – whether a comet or a candle – if you applied a direct force. So they assumed that something must be pushing or pulling the planets round the Sun. Something, they said, like a swirling vortex of magical ether.

But Halley and Hooke rejected this idea. Perhaps, they said, the force that pulls or pushes the planets and comets didn't actually have to touch it. Maybe it could work from afar. They called it 'action at a distance'.

The idea that you could have action at a distance, so that something moved without a point of contact, didn't exist at that time. It's kind of shocking even now. A force acting at a distance, without any intervening medium, is quite spooky.

At the time, most scientists thought action at a distance was evoking the mystical. And so they were very suspicious. But Halley, Hooke and Wren were now thinking the unthinkable. Perhaps, they said, this spooky pushing and pulling somehow emanated from the Sun and somehow sucked the planets and comets to it. And if that was the case, they needed to explore how their spooky planet-moving force changed as it reached across empty space. They wanted to understand how this force fell off with distance. They wanted to understand the mathematics.

Kathy Sykes

About eighteen months later, after a meeting at the Royal Society in early 1684, Halley, Hooke and Wren again fell into a discussion of the orbits of planets. Instinctively they all felt that there was some kind of inverse-square law governing the attraction between the planets and the Sun, but none of them could prove it, not even Hooke, to whom Wren had offered a 40-shilling wager if he could solve the conundrum.

That summer, Halley found himself in Peterborough, most probably on business connected with the settling of his late father's estate. In August he passed through Cambridge and visited Newton, who had also devoted considerable time to observing the comets of 1680/81 and 1682.

According to the story later related by Newton – a story that has become one of the most famous in the history of British science – Halley asked him how the planets would move around the Sun if there was indeed a force of attraction that decreased in proportion with the square of their distance apart.

'It would be an ellipsis,' said Newton.

'But how do you know?' asked Halley.

'Because I have calculated it.'

This was an astonishing boast by Newton. Halley wanted to know more. Immediately.

While Halley watched with anticipation, Newton rummaged among his papers, but couldn't find his notes from his *anni mirabiles*. On Halley's instigation, he promised to rework his mathematical proof and send it to London.

Three months later, in November, Newton sent a nine-page tract, *De motu* (*Concerning Motion*), that provided full mathematical proof of the inverse-square relationship. In the nearly twenty-year period since he had (according to his telling) first undertaken the calculations, more accurate figures for the radius of the Earth had been calculated by the French astronomer Jean Picard. With his improved grasp of calculus and the new data, Newton showed that a large body such as the Earth would act like a single point, attracting the Moon as if it was a conker being spun on the end of a piece of string. More significantly, Newton's mathematical model of the effect of gravity on the Moon's orbit corresponded exactly with the data from experimental observations.

Returning to his old stamping ground of investigative and experimental science had whetted Newton's appetite for more. Over the next eighteen months he wrote *Principia Mathematica* – of which more later – while Halley put up the money to finance its printing, the Royal Society having insufficient funds.

Catalysing Newton into action and acting as midwife to the great man's masterpiece appears to have pushed Halley into taking his vocation seriously again. In the years after the publication of *Principia* in 1687, he devoted his time to drawing up the first charts of trade winds and surveying the east coast of England. This was to rebuff an invasion by William of Orange – unsuccessfully, it turned out; William invaded in November 1688. Asked in 1691 to assist with salvage of gold and ivory from the *Gynie*, a frigate that had sunk off the Sussex coast, Halley invented the diving bell, and then the diving suit. In 1696, he was appointed deputy controller of the country mint at Chester by his old friend Newton, who by then was warden of the Royal Mint, and in 1703 he became Savilian Professor of Geometry at Oxford, a position equivalent to the Lucasian Professor of Mathematics at Cambridge.

In the same period, Halley tackled a variety of subjects in the pages of the *Philosophical Transactions*, ranging across optics, mathematics and rainbows to thermometers and barometers.

An observation that identically sized objects made from different materials had quite different weights led Halley to deduce that matter was more tightly packed in some materials than others, and that led him to investigate the idea of atoms. Although he failed to measure the size of an atom experimentally, he had considerable success with other projects.

From studying mortality records, Halley constructed the first tables of life expectancy for actuaries. By determining the tide patterns in 54 BC, he worked out exactly where Julius Caesar had landed in his invasion of Britain. And through investigations of the biblical flood, assessments of natural erosion and analysis of the saltiness of sea water, Halley revised the accepted 4004 BC date of creation to around 2,000 years earlier, a finding that (although wildly incorrect) led to charges of heresy. Halley believed that his calculation prompted the Church to blackball his application for the post of Savilian Professor of Astronomy when the chair became available in 1691, writing that 'a caveat entered against me, till I can show that I am not guilty of asserting the eternity of the world'.

However, one conundrum from Halley's early days of scientific investigation still intrigued him: comets. With Newton's help, Halley set about recording the trajectories of twenty-four of these dirty snowballs of ice and rock. By the time he was finished, he was convinced that the comet he had observed from Paris in early 1682 was the same comet that had appeared in 1456, 1531 and 1607. The similarity of the path in each case and the interval of 75 or 76 years persuaded Halley that he was dealing with a comet on a very elongated orbit about the Sun and Saturn. In 1705, he published his findings in a book and made the prediction that the same comet would return around 1758 or 1759, depending on the gravitational interference of Saturn.

It was an incredibly bold prophecy, but Halley was proved right when the comet returned sixteen years after his death, first spotted by a German farmer on Christmas Day 1758. Halley's prognosis was correct in all its details, including that Saturn would retard the comet's orbit, and since then the comet has carried his name.

Understanding that comets were as much subjects of the Sun as the planets was only one of several discoveries by Halley that had to wait until after his death for experimental verification. In 1715, when a total eclipse of the Sun threw London into daytime darkness for the first time in hundreds of years, Halley organised a series of systematic observations throughout the country. His records were so exact that they were used again in 1988 to prove there had been no noticeable change in the diameter of the Sun in all that time.

And in a paper based on his observation and timing of a transit of Mercury across the face of the Sun, made during his stay on St Helena at the dawn of his career, Halley showed how observations from several points on Earth of transits of Venus, which he predicted would next occur in 1761 and 1769, could be used to calculate the distance of Earth and Venus from the Sun and the scale of the known solar system.

Nineteen years after Halley's death, the 1761 transit was measured from no fewer than sixty-two observing stations and the distance to the Sun was calculated. Then to verify the findings, in 1766 the Royal Society hired James Cook, a relatively unknown sailing master in the navy with a reputation as a talented surveyor, to travel to the Pacific Ocean to observe the 1769 transit of Venus. Promoted to lieutenant and sailing from England in 1768 on his first voyage as commander of an expedition, Cook rounded Cape Horn and continued westward across the Pacific to arrive at Tahiti on 13 April 1769, where the observations were to be made. Cook's observations were not as accurate as the Royal Society had hoped, but he went on to become the first European to reach the Eastern coast of Australia. Meanwhile, in London, the distance of Earth to the Sun was calculated as 95 million miles, very close to the modern computation of 92.9 million miles.

Although Halley's achievements were overshadowed by Newton (and by Hooke after his recent historical reassessment), he made a remarkable contribution to the emergence of modern science. He spent the last twenty years of his life as Astronomer Royal, re-equipping the observatory on Greenwich Hill with new instruments to replace those seized by Flamsteed's relatives and creditors after the latter's death, and conducting detailed studies of the Moon. However, Halley was much more than solely an astronomer. In the breadth and content of his scientific investigations he rivalled Hooke for creativity and was a far more versatile scientist then suggested by his popular association with comets.

* * *

As one of the few eminent scientists among Newton's peers never to have clashed with the great man, maybe Halley's two greatest achievements were to prompt the great man to return to science after he'd abandoned it for theology and alchemy, and then – by sponsoring the publication of *Principia Mathematica* and seeing it through its printing – to ensure that Newton's epoch-defining discoveries were properly disseminated.

Principia Mathematica was notable not only for transforming science (and in the process, Newton's life and reputation), but also for transforming the way that Newton worked. Previously, Newton had been fascinated by such a wide variety of ideas that he rarely saw any of them through to completion. But *Principia* was different: the concept of a set of universal laws that governed everything was so grand that he pursued each of his theories through until the end.

Working in seclusion for two and a half years, Newton abandoned theology and alchemy to elaborate the ideas he'd touched upon in *De motu*. And the further he explored, the more he discovered that needed further investigation.

Newton had very little contact with the outside world in this period. He gave occasional lectures, which were entirely related to his work on *Principia* and largely incomprehensible to anyone else. He also wrote to other scientists, in particular Flamsteed, whom he asked for information on planetary movements. Newton threw himself into a frenzy of creative thought. What followed is the most incredible three years in scientific history.

Isaac Newton digested the world's astronomical data which the Royal Society had so diligently collated: tidal information brought from Asia by the East India Company; data about eclipses and equinoxes from medieval Arabic astronomers; meteorological information from slave traders and sugar plantation owners in the West Indies; Halley's detailed astronomical readings from his South Atlantic trip and from the Royal Observatory; Boyle's air pump results and Hooke's inverse square law.

Newton was working on a truly global idea. He was convinced that he could explain absolutely everything with one universal law that would answer all their questions, the first Grand Unified Theory of the entire Universe.

Becoming an increasingly eccentric figure, Newton would wander the college with his hair unkempt and his shoelaces untied. His assistant Humphrey Newton (unrelated to Isaac) described a man transported outside himself, who was 'so intent, so serious upon his Studies' that 'he eat very sparingly, nay, ofttimes he has forget to eat at all'. At other times, Newton would leave his room to dine in his college hall, but turn back suddenly to make a correction to his notes, or 'take a Turn or two [in the garden]', suddenly stop, turn around, 'run up ye Stairs, like another Archimedes, with an Eureka, fall to write on his Desk standing, without giving himself the Leasure to draw a chair to sit down in'. In the first few of these intense months of 1685, Newton expanded his ideas in *De motu* regarding the motion of orbiting planets into his three universal laws of motion. The breakthrough came early in 1685 when, drawing on Descartes and Galileo, Newton realised that he could not ignore the principle of inertia and postulated his first law of motion: a body at rest remains at rest, and a body in motion remains in motion at a constant velocity, as long as outside forces are not involved.

In our Newtonian world, the first law of motion might appear obvious, but at a time when many people believed that angels controlled the movement of planets and other bodies, it was a bold statement. More importantly, it had profound implications when combined with Newton's other laws.

For his second law, Newton looked at the effect of outside forces on bodies, a concept he had begun to explore during his *anni mirabiles*.

Like the first law, the second law seems obvious nowadays, particularly to anyone who plays pool or snooker: the change of motion of a body is proportional to the force impressed and happens along the straight line in which that force is impressed.

But the law's simplicity was its beauty. It implied a quantity of matter; a force would change the motion of a small amount of matter to a greater extent than the same force on a larger amount of matter. Newton called this quantity of matter its 'mass' – an entirely new concept. Before Newton, the movement of planets in their orbits, the falling of objects to the ground and the resistance of objects to changes in their motion were regarded as distinct, possibly unrelated phenomena. Newton united these phenomena using the single underlying idea of mass.

In his third law of motion, Newton addressed the equal and opposite changes of motion when two forces impact: for every action there is an equal and opposite action. This law of conservation of motion (which was Newton's name for momentum) explains how jet aircraft and rockets work. If an aircraft's jet engine expels a vast mass of hot air from its exhaust at a very high speed, the equal and opposite consequence will be that the aircraft (a much greater mass than the air) moves in the opposite direction at a somewhat lower speed.

Having derived his three laws of motion, in mid-1685 Newton applied them to the motion of planets. He made pendulums of various different materials – gold, lead, sand, water, wood and wheat, among others – on cords of identical length. Timing the swing of each of the pendulums, he found they all took the same amount of time to complete one oscillation back and forth. It implied that the Earth was attracting the various substances in exact proportion to the amount of matter in the substance, in other words, in proportion to their mass.

Using his three laws of motion, Newton then worked out the way in which the gravitational force between the Earth and the Moon could be calculated. He found that it was proportional to the product of the two masses [m_1 and m_2] and inversely proportional to the square of the distance [r] between their centres.

Newton then introduced a constant of proportionality called the gravitational constant or G, to create an equation that expressed his law of universal gravitation in mathematical terms. The equation is one of the most famous in science and certainly the most fundamental expression of the forces at work in the universe until Einstein investigated relativity.

$$F = G\frac{m_1 m_2}{r^2}$$

Although this equation and those describing Newton's laws of motion were at the heart of *Principia Mathematica*, they begged many other questions and explanations, which Newton expanded into his epic three-book masterpiece. They explained all of Kepler's laws of planetary motion, including various irregularities, the precession of the equinoxes, the pattern of tides, the bulge of Earth at its equator, and several other astronomical and terrestrial phenomena.

By April 1686, the first book was finished and Newton sent a manuscript to the Royal Society.

With his new mathematical theory, his laws of motion and his iconoclastic idea about gravity, Newton set about explaining the universe. And everything fell into place. The *Principia* completely explained the elliptical orbits of the planets. It explained that the Moon was speeding up because of gravity's effect on the tides here on Earth. It even explained the comet.

The comet followed the same laws as the larger planets, but its orbit was so long and thin that it appeared to travel in straight lines across the sky, disappearing and reappearing at intervals.

Newton had unearthed a deep and fundamental truth about the world: that there are universal laws governing how things move, that there is an invisible force which connects and attracts all objects around us, and that mathematics is the key to all scientific explanations.

It was the grand culmination of the experimental work of Hooke, Boyle and Halley. Newton brought it all together.

Jim al-Khalili

Three weeks later, the Royal Society decided to publish *Principia*. Another two days later Halley wrote to Newton. Assuring Newton that his 'incomparable treatise' would be published, Halley added that Hooke had seen the manuscript and raised the cry of plagiarism. Given that the inverse-square relationship had been discussed widely, but only Newton had been able to prove it mathematically or link it via mass to a universal law of gravitation, it is understandable that Hooke's accusation sent Newton into a flaming rage. 'Now is this not very fine?' Newton wrote. 'Mathematicians that find out, settle and do all the business must content themselves with being nothing but dry calculators and drudges. And another that does nothing but pretend and grasp at all things must carry away all the invention.'

Fearing a controversy, Samuel Pepys, at that time the president of the deeply indebted Royal Society, reneged on his decision to publish. Halley immediately stepped in, offering to pay all costs, including for illustrations, and committing to edit the book himself.

Newton was not placated. He removed a generous acknowledgement of his debt to Hooke from volume three of *Principia* and wrote to Halley, threatening to withdraw the third book from publication. For Halley, Newton's threat spelt financial disaster, but he mollified Newton and suppressed Hooke sufficiently for Newton not to mutilate his masterwork. On 5 July 1687, it appeared in print in an edition of 2,500 copies.

Principia Mathematica was immediately recognised as the most important book published about science, if not about any subject – a feat that has not been surpassed to this day. Other scientists, such as Huygens, travelled to England solely to meet Newton, awed that for the first time the entire universe had been revealed as working on basic mechanical principles that were not mystical or spiritual and that were understandable by anyone.

In making the universe a matter of measurement and mathematical equations, Newton had overturned thousands of years of classical philosophy with a series of laws that were far more elegant and revealing than anything derived by the Ancients. *Principia Mathematica* was the logical culmination of the scientific revolution that had begun with Copernicus and been properly formalised by Gilbert. Science had come of age. The Age of Mysticism had been replaced by the Age of Reason.

After publishing *Principia*, Newton emerged from his shell for the first time since he had retreated from public life eleven years earlier following the criticisms of his work on optics and mathematics published in the *Philosophical Transactions*. He sat for portraits, met other notable mathematicians and philosophers, including John Locke, who became one of his closest friends and one of the few people with whom Newton felt he could discuss his religious views. He attended meetings at the Royal Society, as well as the funeral of Robert Boyle. He also appeared in court to oppose the imposition of Catholic influence on the University of Cambridge and subsequently became one of two Members of Parliament for the university.

Although Newton had switched most of his intellectual energy to alchemy, he sought to capitalise on *Principia*'s success by working on a second edition in which he intended to unite his ideas on calculus and optics.

Then in 1693 he suffered a nervous breakdown. In letters to John Locke and Samuel Pepys, his closest friends, he described a 'discomposure in head or mind or both'.

Years of overwork and living a secretive life had taken their toll. Newton's legendary paranoia and sensitivities were as much the result of concealing his unorthodox Arian religious beliefs as wanting to keep details of his work private. But possibly the biggest cause of his breakdown was the sudden ending of his relationship with a brilliant young Swiss mathematician called Nicolas Fatio de Duillier. Having never married or expressed interest in relationships with women, and having conducted several intense relationships with unmarried men, Newton has been assumed by many historians to have been homosexual. His relationship with Fatio, described as 'incandescent in its emotional intensity', ended at about the same time as his nervous breakdown, but it is not known whether the end of the relationship prompted the breakdown, or vice versa, if at all.

When he recovered, Newton realised that his creative days were over. He immediately began to search for an escape from Cambridge, finding it in 1696 in an offer of the Wardenship of the Royal Mint, which needed reform to counteract widespread counterfeiting.

Newton made a great success of the Mint, instigating a recoinage to counteract the scourge of clippers who, by successively scraping away small amounts of precious metal from the edges of silver and gold coins, had reduced most of Britain's coinage to half its original size and weight. Newton introduced a mechanical minting process to produce coins with milled and inscribed edges so that clipping could be detected. He even became a magistrate to ensure that counterfeiters were convicted and hanged, but it kept him from finishing his second edition of *Principia*, although he did find time to conduct one of his characteristic vendettas, this time against Flamsteed, with whom he clashed over his map of the stars.

In 1703, when Hooke died, Newton accepted election to President of the Royal Society and the next year published *Opticks*, a summary of his work on colour and light conducted nearly forty years earlier, but sat on until Hooke was no longer capable of voicing dissent.

Newton made as much of a success of the Royal Society as he had previously achieved with the Mint. He increased membership, attracted crowds to its meetings and put it on a sound financial footing that culminated in a move to new premises during which the only known portrait of Hooke mysteriously disappeared.

In 1705, Newton became the first scientist to be knighted, although the honour was more for his work at the Mint than his great discoveries. Meanwhile his dispute with Leibniz over priority in the invention of calculus, which dated back to 1684, rumbled on, coming to a head in 1712 when a committee was established by the Royal Society to examine the dispute. The next year, Newton finally published a second edition of *Principia*, aided by a young Cambridge mathematician, Roger Cotes, who reignited Newton's long-extinguished passion for science. Thanks to Cotes's painstaking efforts, several new ideas were added and other theories revised that had not been fully realised in the first edition. Three years later, Cotes died suddenly of a violent fever, aged 33. Although he clashed with Coates during the revision of *Principia*, Newton appears to have had a rare respect for the young man, regarding him as one of the few mathematicians capable of following in his footsteps and commenting after his death that 'if he had lived, we would have known something'.

The dispute with Leibniz was never settled; Leibniz died in 1716, but his supporters continued their argument in philosophy and literature journals for another six years until interest died down.

In his final years, the focus of Newton's philosophical endeavours returned to theology and he made few new contributions to science beyond attending to the publication of new editions of *Opticks* and *Principia*, both of which had sold out and been extensively pirated in counterfeit editions.

Newton's laws of motion, his theory of gravity and his mathematics were the pinnacle of human thought for the next two centuries.

They enabled us to put a man on the Moon and me into space. They allow us to reach out into the universe to explain its intricate mechanism.

But although he was the greatest genius of them all, Newton was indeed only standing on the shoulders of giants. And among those giants were Robert Hooke, Robert Boyle, Edmond Halley and Christopher Wren. Together in the seventeenth century they summoned science into being.

Stephen Hawking

After his death in March 1727, Newton's body lay in state in Westminster Abbey before a burial in the nave beneath an extravagant monument inscribed: 'Let Mortals rejoice that there has existed such and so great an Ornament to the Human Race.' The poet Alexander Pope was moved by Newton's accomplishments to write the famous epitaph: 'Nature and nature's laws lay hid in night; God said "Let Newton be" and all was light.'

However, an uncharacteristically modest description of his life, supposedly written by Newton, serves maybe as a more fitting eulogy: 'I do not know what I may appear to the world, but to myself I seem to have been only a boy playing on the sea-shore, and diverting myself in now and then finding a smoother pebble or a prettier shell than ordinary, whilst the great ocean of truth lay all undiscovered before me.'

Chapter 6

CHEMISTS SPARK
THEIR OWN REVOLUTION

Newtonian physics had a profound effect not only on science, but also on the entire culture of Britain. In science, Newton's explanation of the physical world in terms of mathematical equations and laws encouraged biologists and chemists to apply the same rational and mechanistic approach to their fields. Robert Boyle had done much to change perceptions of previously mysterious substances – gases were now known to have tangible properties – but chemistry was still struggling to escape its mystical roots in alchemy. In part this was because of an intellectual reluctance to apply rationality to the investigation of substances and their properties. However, an equally significant factor was the difficulty of designing instruments and methods necessary to investigate chemical properties in a rational, experimental manner.

In the wider culture, the principles of experimental science – induction over deduction; rationalism over mysticism – were shaping the Enlightenment. The Royal Society was more active than any other British institution in spreading Enlightenment ideas around Europe, in particular the experimental philosophy of Boyle, Hooke, Halley and Newton. So it is perhaps not such a coincidence that Newton's *Principia*, the chronological and intellectual climax of the scientific revolution, was published in 1687, the year directly before the Glorious Revolution, often cited as the beginning of the Enlightenment in Britain.

The bloodless revolution of 1688 that saw James II overthrown by a union of Parliamentarians in cahoots with an invading army led by William III of Orange came at a time of profound change in Western philosophy and cultural life. Mysticism, theocracy and the divine right of kings were giving way to rationalism, self-governance and modernism. The rational principles at the heart of experimental science – and which were best exemplified by Newton's law of universal gravitation – were being adopted in all areas of intellectual thought, in the arts, humanities and in philosophy generally.

But for all its elegance as a grand unified theory of the universe, Newton's law of universal gravitation had an Achilles' heel: no one knew the precise value of G, the gravitational constant. Without G, it was impossible to measure the exact gravitational attraction between two bodies.

The only way to calculate G was to measure the gravitational force between two bodies of known masses, but that was easier said than done. Gravitational force between Earth and most bodies is very weak, which explains why, for example, we can pick up a dropped teaspoon so easily. It only becomes large when the two bodies are massive, such as the Moon orbiting Earth. However, it's very difficult to perform laboratory experiments on objects as large as planets and their satellites.

Nevertheless, Henry Cavendish, a highly eccentric and very shy recluse, provided a way of calculating G more than 100 years after Newton formulated his law. Cavendish came closer than any other major scientist to embodying the caricature of a mad boffin and he did so by devising a suitably madcap experiment. He weighed the world. On, of all places, Clapham Common.

Cavendish was one of the richest men in eighteenth-century Britain. The Duke of Devonshire was his paternal grandfather, the Duke of Kent his maternal grandfather, and he owned the single largest holding of bank stock in England. Born in 1731 on the French Riviera (where his mother had been spending time on a convalescent holiday), he was destined for a life in politics and, after schooling in London, he arrived at Peterhouse, Cambridge, in 1749. However, he never completed his degree, partly because he objected to the religious requirements of his education, but mainly because he was too shy to face the professors in his final examination.

Cavendish's shyness was so acute that modern psychologists have suggested that he suffered from Asperger's syndrome. If approached by a stranger, he would flee. On the rare occasions that he spoke to an acquaintance, a high-pitched stammer would invariably preface a rustling sound as Cavendish bustled away quickly. Conversing with more than one man at a time was impossible for him; likewise talking to any woman at all. To avoid meeting his female housekeeper, Cavendish had a back staircase added to his house. When required to communicate with his cook, he would do so only via written notes, usually an instruction for dinner: 'a leg of mutton', or on the rare occasions he was expecting guests, 'get two'.

Cavendish built a concealed entrance to his house so that he could arrive and leave without anyone knowing, and moved his library to a separate house four miles from his home, so that he wouldn't have to meet anyone who wanted to borrow one of his books. Given his personality, it's unsurprising that Cavendish did not follow his ancestors into Parliament. Instead, he devoted himself to his only true passion: science.

> Cavendish was so intensely shy that he always refused to have his formal portrait painted. Only one picture of him exists. And it was made without his knowledge or permission.
>
> The artist first drew Cavendish's hat and frockcoat hanging in the cloakroom at the Royal Society. He then snatched a profile of Cavendish sitting in the dining room eating supper. And then he put all the pieces together to make the portrait.
>
> Jim al-Khalili

Having converted large parts of his large house on Clapham Common into laboratories for studying chemistry and physics, Cavendish would shirk no challenge if it promised enlightenment. In one experiment to measure the strength of electrical current, he subjected himself to increasingly severe electric shocks, noting the degrees of agony until he passed out. In another experiment, he blew a mouthful of hydrogen across an open flame, the loss of his eyebrows providing painful proof that hydrogen is indeed highly flammable.

The great tragedy of Cavendish's life – more than his crippling shyness – was that he kept details of most of his discoveries completely secret; he made Newton look like a blabbermouth. As a result, the names of scores of other scientists are associated with discoveries that Cavendish had made tens of years earlier.

In fact, Cavendish achieved his reputation as one of the giants of modern science only many years after his death, when the Cambridge physicist James Clerk Maxwell found in Cavendish's papers evidence that he'd made discoveries in electricity that were fifty years ahead of his time.

Ohm's law of resistance, Dalton's law of partial pressures, Charles's law of gases, the law of conservation of energy, Richter's law of reciprocal proportions, Coulomb's law of electromagnetism – all of them might have carried Cavendish's name had he not been so tight-lipped about his work.

Cavendish embarked on his attempt to weigh the world in the twilight of his long career. In the summer of 1797, aged 67, he assembled a machine in a shed in the garden of his Clapham Common house. The machine, designed by John Michell, a geologist and friend who had died before he could use it, resembled nothing so much as a giant executive desk toy. Two small balls were suspended from a horizontal rod attached to the ceiling by a single wire. Two larger lead balls were placed nearby.

Cavendish knew the exact masses of each set of balls and the precise distances between their centres. He also knew the force of attraction between the balls and the Earth; in other words, their weights. So if he could measure the force of attraction between the sets of balls in his contraption, he could infer the weight of the world. From these figures, he could also work out the gravitational constant, G, although he left the details of that calculation to another scientist.

Knowing that the gravitational attraction between the balls would be astonishingly small (in fact, it turned out to be five hundred million times less than the weight of the lead balls), Cavendish's biggest challenge was to eliminate every source of error that could disrupt the precision of his experiment.

Working at this featherweight scale, the mass of Cavendish's own body would have generated more gravity than that between the sets of balls, so he removed himself from his shed. More than likely dressed in a faded velvet coat and a three-cornered cocked hat that had last been fashionable in the previous century (Cavendish had the same suit remade whenever it wore out and, despite his great wealth, only owned one suit at a time), he peered down a telescope eyepiece into his shed. In a series of seventeen experiments that took him nearly a year, Cavendish painstakingly measured the torsion (the amount of twist) in the wire holding the smaller spheres. He then reported the result of his calculations in the *Philosophical Transactions*.

Strictly speaking, Cavendish reported his result in terms of the density of Earth, which he said was 5.48 times the density of water. But in modern weight terms, the world was found to weigh 6.6 billion trillion metric tons, or 6,600 followed by 18 zeros. Amazingly, Cavendish's result was within 1 per cent of the current measurement of Earth's density and only marginally different from Newton's educated guess a century earlier.

Although Cavendish is best known for weighing the world – the feat is remembered as the Cavendish experiment and the Cavendish Physical Laboratory is named in his honour – his other experiments with electricity and gases were more significant developments in the history of physics and chemistry. These began in 1766 with his paper on 'factitious airs', as gases produced by chemical treatment of solids or liquids were known then.

Gases were deeply mysterious, barely understood entities in the eighteenth century. Even the word 'gas' was new, having been coined when a heavily accented Dutch chemist, J. B. Van Helmont, was misheard when he used the Greek word 'chaos' to describe the nature of air.

Cavendish concocted several experiments to produce gases, which he then examined, the most notable experiment involving dissolving metals in acids to produce hydrogen, which he called inflammable air. Boyle had previously isolated hydrogen, but Cavendish was the first to determine that this inflammable air was an element in its own right, quite different to the air we breathe and also highly flammable, as his singed eyebrows could attest. He then went on to investigate the various properties of hydrogen, including its density and solubility.

While looking into what happened when metals dissolved in acid, Cavendish discovered hydrogen. It's the simplest atom there is, it's the fuel that powers the sun and stars. Ninety-five per cent of everything in the universe is made of it.

ROGERIVS BACO,
Monachus in Anglia
Aſtrologiae Chemiae et Mathe,
ſeos peritiſſimus.
Nat: A.1206.   Don. Aizee,
Ex collectione Friderici Roth-Scholzii Norib.

The three godfathers of British science. The Venerable Bede (bottom right) computed the date of Easter and wrote the first British scientific text. Roger Bacon (left) was the first to advocate a 'science of experience' – the use of repeatable experiments to verify scientific theories. Regarded by peers such as Galileo as the world's first true scientist, William Gilbert (bottom left) established the principle of the scientific method – nothing accepted as fact unless proven by extensive observations from repeatable experiments.

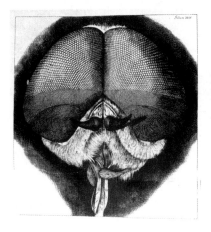

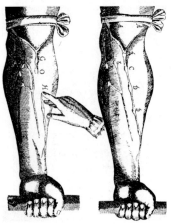

The greatest experimentalist of his generation, Robert Hooke perfected the compound microscope and opened up the realm of the very small to the gaze of the world in his book, *Micrographia*.

William Harvey's discovery of the workings of the circulatory system brought scientific principle to medicine and began its liberation from the orthodox Galenic theory of humours of the body.

After Robert Boyle used an air pump to investigate the properties of air, showmen would travel the country demonstrating his discoveries at public meetings, as seen in this painting by Joseph Wright.

ISAACVS NEWTONVS.

Dismissive of others, obsessive, egomaniacal and paranoid – Isaac Newton was all of these, but he was also an unparalleled genius, his discovery of the first grand unified theory of the universe making him the greatest scientist of all.

A model of the shed on
Clapham Common in
which Henry Cavendish
used an apparatus not
dissimilar to a giant
executive toy to weigh
the world, thereby
proving Newton's law
of universal gravitation.

So shy that he was incapable of speaking to his
female housekeeper, Cavendish made many
groundbreaking discoveries later attributed to
other scientists, but failed to publicise them.

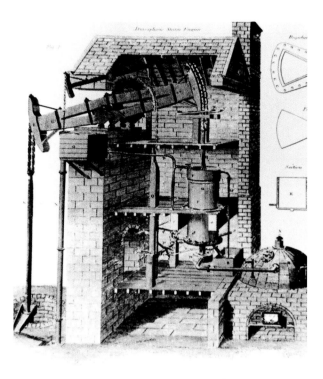

Thomas Newcomen developed the first working steam engine, although his design relied on the weight of the atmosphere to do the work and was therefore strictly speaking an atmospheric engine.

From observations of his experiments with steam models, James Watt struck upon the idea of using a separate steam condenser to develop the first efficient steam engine. It changed the course of history.

Humphry Davy's lectures at the Royal Institution were the talk of fashionable London, as much because of his charisma and good looks as his discoveries. However, his celebrity led his discovery of several gases – including hallucinogenic nitrous oxide – to be widely ridiculed by cartoonists.

With his high pressure steam engine built onto a stagecoach, Richard Trevithick wanted to launch the world's first steam-powered passenger vehicle service. He drove his London Steam Carriage from Holborn to Paddington in London, but it proved uncomfortable and more expensive to run than horses and failed to catch on.

The feet behind John Hunter in this painting by Joshua Reynolds
of the great anatomist and pioneering surgeon belong to the 'Irish
giant', Charles Byrne. Hunter desperately wanted him for his unique
collection of animals and humans, now on display at the Hunterian
Museum in London.

The Cow Pock — or — the Wonderful Effects of the New Inoculation! — Vide. the Publications of ye Anti Vaccine Society

A caricature of a woman being inoculated by
Edward Jenner, who coined the word vaccination to
describe his use of cowpox inoculation to obtain
immunity to smallpox. Within years of his discovery,
smallpox vaccination was widespread and by 1979
the disease was eradicated.

Entirely self taught, Michael Faraday came relatively late to
science but arrived in spectacular fashion, soon surpassing his
mentor Humphry Davy and discovering the fundamental
relationship between electricity and magnetism, thereby
triggering a slew of inventions – the electric motor, dynamo
and telegraph – that changed the world. He is seen here in a
stereoscopic daguerreotype, taken less than a year after the
birth of photography.

Cavendish filled a sheep's bladder with his new gas and weighed it. He found it was about 11 times lighter than ordinary air. This suggested exciting possibilities.

Soon the first hydrogen balloon was rising into the sky, and then came the first aerial crossing of the English Channel. For the first time, people could see the earth from above.

Later there would be huge airships crossing the Atlantic. And later still the H-bomb. And soon, we hope, clean fusion power to generate electricity, and hydrogen-powered cars to help save the planet from the effects of climate chaos. All thanks to the gas discovered by Cavendish.

Jim al-Khalili

Most chemists experimenting with gases at that time subscribed to the phlogiston theory to explain why certain substances would burn. This put forward the idea that in addition to the four classical elements of the Greeks (air, fire, water and earth), combustible materials contained a colourless, odourless, tasteless and weightless material called phlogiston. When a combustible substance was burned, phlogiston was liberated, leaving the dephlogisticated substance as a residue. The extreme flammability of hydrogen and its very low weight led Cavendish to believe that he had isolated the mystical substance, phlogiston.

At about the same time, Joseph Priestley, a radical nonconformist preacher and good friend of Cavendish, had discovered another gas that he called dephlogisticated air because, although it would not burn, it was found to be necessary for combustion to take place and seemed to absorb phlogiston. The explanation made absolute sense, as the gas was in fact oxygen.

In 1783 Cavendish invented a new machine, which he called a 'eudiometer for measurement of the goodness of gases for breathing'. It had a sealed vessel that allowed him to burn precisely measured amounts of gases and analyse their products. Cavendish discovered that when he burnt his inflammable gas in Priestley's dephlogisticated air, the two gases combined explosively to make water – quite a blow for adherents of the classical Greek view that water was an immutable element, particularly as the weight of the water was the same as the combined weights of the two gases. Through this experiment, Cavendish proved that water was a compound made up of the two elements, and that two gases could combine to make a liquid, a discovery that was again counter to the classical Greek view. As a result, by 1787 Cavendish had abandoned the phlogiston theory in favour of a new antiphlogistic theory developed by a French chemist, Antoine Lavoisier.

Intrigued by his eudiometer experiments, Cavendish tried burning varying amounts of hydrogen in different amounts of fresh air, which he called common air. From these experiments he discovered that to make water he needed twice the volume of hydrogen as oxygen – hence H_2O. He also discovered that using up all the oxygen left another gas that wouldn't support hydrogen combustion. He called this gas phlogisticated air; we now know it as nitrogen.

In a further experiment in which he passed electric sparks through various mixtures of nitrogen and oxygen, he eventually discovered the composition of atmospheric air: roughly 79 per cent nitrogen, 21 per cent oxygen and a tiny amount of a gas that he couldn't identify. The findings were remarkably accurate and a testament to the incredible precision of Cavendish's methods and equipment. Cavendish dismissed the small quantity of unidentifiable gas as probably no more than an experimental error, but more than a century later he was vindicated when the unidentified component was discovered to be the rare gas argon, which makes up 0.93 per cent of the atmosphere.

* * *

Cavendish was only one in a line of so-called pneumatic chemists who investigated gases in the eighteenth and nineteenth centuries. The inspiration for their exploration was a curiosity about the nature and purpose of air. Common sense, as well as Harvey's discovery of the workings of the circulatory system and Willis's painstaking dissections of the brain and nervous systems, indicated that air was somehow vital to life, but beyond that no one knew how or why it was so essential. The first of these pneumatic chemists, whose work on gases predated Cavendish by about fifteen years, was a leading member of the Scottish Enlightenment movement called Joseph Black.

If Boyle was the father of chemistry who used experimentalism to challenge the mystical aspects of alchemy, then Black was its midwife for applying precise quantitative techniques that had been pioneered in physics to the new science of chemistry.

CHEMISTS SPARK THEIR OWN REVOLUTION

Born in France to a Scottish wine merchant working in Bordeaux, Black was pressed by his father to study medicine. Black was then led by his curiosity about kidney stones to write a doctoral thesis for his medical degree that contained fundamental insights into the nature of air. It became a classic in the history of chemistry.

In the mid-eighteenth century, the common treatment for conditions such as kidney stones or an acid stomach was to prescribe a remedy that often contained alarmingly caustic salts. White magnesia, which we now know as magnesium carbonate, was given for acid stomachs, so Black decided to investigate if it would also work for kidney stones. It didn't. But in the process Black discovered that if he heated white magnesium strongly it gave off a gas. No steam or liquid vapour was produced in the course of heating, so Black surmised that all the gas had come from the white magnesia and that the weight of the gas was the same as the weight lost by the powder when he heated it.

Intrigued by his finding, Black then looked at other similarly caustic salts, such as limestone (calcium carbonate) and discovered that not only did they all give off the same gas in the same way when heated, but that when he added acid to them, they all fizzed and produced exactly the same gas in the same proportion as when he heated them.

Black decided to call this gas fixed air because it appeared to be 'fixed' to the salt until the salt was heated or dropped into acid. The name also arose because the gas could be 'fixed' back into the heated salt if he dissolved it in water and then added a mild alkali.

This fixed air, Black subsequently discovered, would extinguish the flame of a candle; we now know it as carbon dioxide.

Black also noticed that quicklime (calcium oxide) could be converted back to limestone (calcium carbonate) simply by leaving it exposed to air. It implied that the atmosphere contained fixed air and provided the first proof that the atmosphere was a mixture of gases, a direct contradiction of the classical model, in which air was a fundamental indivisible element of nature.

In the burgeoning field of chemistry, Black's discoveries established two important principles. By involving a gas in a chemical reaction, he'd shown gases were chemically no different from solids or liquids. And by meticulously weighing every reagent and product, he had founded quantitative chemistry, a field that would become increasingly important over the next century as industrialists looked for efficient ways to exploit the discoveries of science. All these findings were summarised in Black's thesis, *De humore acido a cibis orto, et magnesia alba* (*On Acid Humor Arising from Foods and on White Magnesia*). Its publication turned Black into a celebrity of the scientific world.

Within a couple of years Black was professor of medicine and lecturer in chemistry at Glasgow University. With a private medical practice on the side, he had a reputation for fascinating lectures that drew students from around the world. In part, those students were attracted by the fact that Black had given up publishing the results of his latest research, preferring instead to divulge his new discoveries at his lectures. Consequently, his students were the first to hear his finding that fixed gas was present in exhaled breath and that it was also produced by fermentation. Some of Black's more enterprising students collated their lecture notes and published them in a very successful book.

By his early thirties, Black had moved on from chemistry to physics. Many of his students came from families of whisky distillers and they asked why distilleries needed to expend so much wood (and time and cost) to distil and condense the liquor. Black investigated and in 1764 became the first person to discover the difference between quantity and intensity in heat. Black realised the intensity of heat given off was a measure of its temperature, but the quantity of heat in a body depended on other factors, such as whether it was a liquid, gas or solid. When water reached its boiling point, Black discovered that no matter how much heat he applied, it did not get warmer. The same occurred when he applied heat to melting ice; it did not increase in temperature. Instead it melted, increasing in heat only when it had all turned to liquid water, or in the case of boiling water, it increased in heat only when all the water had been turned into steam vapour. Black called the heat required to change the physical state from solid ice into liquid water, or from liquid water into steam vapour, latent heat. He went on to discover that when the reverse occurred, such as steam condensing to water, or water freezing to ice, latent heat was released.

These early discoveries in thermodynamics were of profound importance to the pioneers of the Industrial Revolution, particularly a young technician at Glasgow University called James Watt, of whom more later.

* * *

Another member of the university to profit from Black's research was one of his young medical students called Daniel Rutherford, an ingenious, gentlemanly young man who joined the ranks of the pneumatic chemists when Black passed on some unfinished parts of one of his research projects.

Black had burned a candle in a closed container until it was extinguished (at the time he didn't know that this was because the oxygen had been used up). When he then used chemicals to absorb all of the carbon dioxide, a gas remained that would still not support combustion, but which was obviously not carbon dioxide. Puzzled, he turned the problem over to Rutherford.

Rutherford devised an experiment in which he confined a mouse in a sealed chamber until it died. He then burned a candle followed by some phosphorus in the gas from the chamber, then passed the remaining gas through a strong alkali to absorb any traces of fixed gas (carbon dioxide). Having used respiration (the mouse), combustion and chemical removal to eliminate gases, Rutherford had effectively gone a stage beyond Black and therefore was confident he had isolated a new gas. Declaring that it would not support burning or allow breathing because it was fully saturated with phlogiston, Rutherford declared that the gas was 'phlogisticated air', the gas discovered earlier by Cavendish, but because of Cavendish's reluctance to trumpet his achievements, previously unknown to the rest of the world. In fact, as we've already noted, it was nitrogen, but Rutherford was given credit for its discovery, although it took other scientists, such as the French chemist Lavoisier, to describe its properties in full.

* * *

With phlogisticated air and fixed air identified, logic suggested that some kind of dephlogisticated air remained to be discovered. According to the phlogiston theory, this would be the gas that supported combustion because it was devoid of phlogiston and eager to absorb it from any burning material.

In spite of the erroneousness of the phlogiston theory, there was logic in the inference and the race was on to discover the atmosphere's missing member. The honour went to Joseph Priestley, who, as mentioned earlier, was a close friend of Cavendish.

Like Newton and Hooke, Priestley had been an extremely bright child with a troubled and unconventional upbringing that played no small part in shaping his free-thinking attitudes. Born in 1733 to a Dissenting family of weavers in West Yorkshire, he was sent to live with his grandparents at the age of one. His precociousness soon became apparent; at four he could recite flawlessly all 107 questions and answers of the Westminster Shorter Catechism. He returned to live with his widowed father at six, but only until his father remarried two years later, after which Joseph left for his wealthy aunt and uncle's Calvinist home. The stress associated with a lengthy bout of tuberculosis in his teenage years left him with a permanent stutter, while the disease itself fostered a scepticism about his Calvinist upbringing, but neither had any effect on his education. By the time he left school he had mastered Greek, Latin, Hebrew, French, German, Chaldean, Syrian and Arabic, and was preparing for a life serving the Church.

At Daventry Academy, Priestley was considered so far ahead of his classmates that he was allowed to skip the first two years of his theology studies. By the time he matriculated, he was a Rational Dissenter, a branch of Anglicanism that advocated rationality over dogma and religious mysticism. Influenced by the writing of the English philosopher David Hartley, Priestley subscribed to a view that religious and moral issues could be scientifically examined and proven.

For twelve years after leaving Daventry, Priestley worked as a minister at Nantwich in Cheshire, frequently clashing with his parishioners' beliefs, although their sympathy with his stammer enabled him to get the impediment under a greater degree of control. While at Nantwich, he opened a small school with two rooms in which he taught boys and girls, separately, English, Latin, Greek, history, geography, mathematics, natural history and natural philosophy. The success of the school led to an appointment as a tutor at Warrington's Dissenting Academy, where he taught languages and literature. While at Warrington, he published a book on grammar, a seminal work that established him as one of the great grammarians of his time. It was the first in a series of publications covering education, the history of science and Christianity, several of which were important texts in their fields, popular for decades and adopted by schools and universities in Britain and America.

In 1765, Priestley received an honorary Doctor of Law degree from the University of Edinburgh in recognition of his most recent publication, *A Chart of Biography*. Produced in graphical form, it was designed by Priestley to 'impress' upon students 'a just image of the rise, progress, extent, duration, and contemporary state of all the considerable empires that have ever existed in the world'. It was his most successful publication to date.

Shortly after receiving his degree, Priestley made one of his frequent visits to London. One evening he was introduced to Benjamin Franklin – a budding author, inventor, politician, scientist and founding father of the United States of America – and several other 'electricians', as the band of scientists interested in electricity called themselves.

Inspired by the work of William Gilbert nearly two centuries earlier, these electricians studied charges and static electricity, usually generated by friction using an electrostatic machine or simply by rubbing a suitable material, such as amber or glass with silk or wool.

Franklin had become so fascinated by electricity that he had sold many of his possessions to fund his work and had flown a kite in a thunderstorm to find out if lightning was electrical in nature. With a metal key attached to the bottom of a dampened silk kite string, Franklin had watched as sparks jumped from the key to the back of his hand. This reckless experiment proved that lightning was indeed a form of electricity, although it killed the next two people who tried it.

Impressed by Franklin's work, if not his experimental methods, Priestley decided his next book would be a history of electricity.

He returned to Warrington, which was gaining a reputation at that time as the Athens of the North on account of the quality of the education at its Academy and the number of locals involved in debating natural philosophy. Writing and experimenting in the spare time from his teaching obligations, and entirely self-taught in science, Priestley made several discoveries, including that carbon conducted electricity, thereby overturning the belief that only water and metals were conductors.

Priestley also discovered that an electrostatic charge is concentrated on the outer surface of a charged body and that there is no electric force inside a charged hollow sphere, a finding that led him to propose an inverse-square law for charges, analogous to Newton's universal law of gravitation.

Within a year of starting his scientific career, Priestley's achievements had earned him a Fellowship of the Royal Society; within two years he had written his 250,000-word history of electricity, which became the standard text on the subject for the next century. He followed it up within a year with a version for the general public titled *A Familiar Introduction to the Study of Electricity*, becoming probably the first in a long line of popular science writers. His best work, however, still lay ahead of him.

In the year that he published his history of electricity, Priestley moved to Leeds, where he took on the pastorate of Mill Hill Chapel, fortuitously next door to a brewery. Although Priestley had little interest in the brewery's products, he was intrigued by the gas given off by its fermenting brews, which he knew to be Joseph Black's fixed air. By extinguishing a burning candle in the layer of fixed air, he added smoke to the gas and then watched as it spilt over the side of the brewing vat onto the ground. Fixed air was clearly heavier than normal atmospheric air.

A few years earlier, Joseph Black had shown that atmospheric air was a mixture of gases including fixed air. Knowing that fixed air sustained neither life nor combustion, Priestley experimented by once again trapping a mouse in a jar until it lost consciousness, then looked for ways in which he could somehow revitalise the air. After much experimentation, he found that if he put a mint plant in the jar, the mouse regained consciousness. Priestley believed he had returned the phlogisticated air (carbon dioxide) to its dephlogisticated state (oxygen) in which it could sustain respiration. Although he got the terminology and the method wrong, the principle was correct. It was the first indication of photosynthesis in plants, that magical process by which plants use sunlight to turn carbon dioxide and water into food and oxygen.

Next, Priestley dissolved fixed air in some water, pouring the water repeatedly between two jars containing the gas. Then, in the true spirit of curious experimentalism, he took a sip. To his surprise, he discovered the fluid was a refreshing sparkling drink. Priestley had invented soda water, although it took a German watchmaker called Johann Jacob Schweppe to realise the commercial potential of Priestley's discovery and then to develop a process to manufacture carbonated water.

The discovery played a part in Priestley winning the Royal Society's Copley Medal. The award was given not so much for inventing fizzy pop *per se* but because, while later attempting to find a cure for scurvy for James Cook's second voyage to the South Seas, he improved on his method and impressed the Royal Society with the discoveries and experimental techniques he made in the process. In this case, he generated fixed gas through dissolving chalk in sulphuric acid. He then dissolved the fixed gas into water under pressure to produce a much fizzier drink.

Priestley's invention of soda water didn't solve the problem of scurvy at sea but it triggered a European craze for carbonated drinks, which brought him to the attention of William Petty.

Petty was better know as the second Earl of Shelburne, a young Whig parliamentarian who as Secretary of State for the Southern Department was responsible for negotiating with the American colonies shortly before the American War of Independence. Intrigued as much by Priestley's well-publicised criticisms of British attitudes towards colonialists as by his scientific achievements, Lord Shelburne hired him, ostensibly as his librarian but in practice as his political adviser and tutor to his children.

With a salary of £250 a year and free accommodation on Shelburne's estate, Priestley had plenty of time to pursue his science. Over the next few years, he isolated and studied ten gases, including ammonia, sulphur dioxide, carbon monoxide, hydrogen chloride and nitrous oxide. But the discovery that made his name happened shortly after he received a thick glass lens. Eager to experiment with the lens, he picked up some brick-red mercury oxide powder that he'd made earlier by heating mercury in air. Focusing the Sun's rays through the lens onto the red dust, which he called calx, he watched as the Sun's intense radiation returned the calx to shining globules of mercury, driving off an intriguing gas in the process.

This gas, Priestley discovered, had all the properties of atmospheric air, but better. A flame burned more brightly and for longer in it. Mice were particularly frisky when they breathed it and they survived for much longer in a closed container containing the gas than in normal air. Declaring the gas to be 'five or six times better than common air for the purpose of respiration, inflammation, and, I believe, every other use of common atmospherical air', Priestley reported that he had discovered dephlogisticated air. It was the secret of aerobic life, the gas we now call oxygen.

In tests on himself, Priestley said he felt 'light and easy' if he breathed it and, anticipating the oxygen bars of Japan and California of the late 1990s, predicted that it might become a rich person's indulgence one day, although 'hitherto only two mice and myself have had the privilege of breathing it'.

Although the discovery of oxygen was attributed to Priestley, the Swedish chemist Carl Scheele and his French counterpart Antoine Lavoisier also laid claims to its discovery. In retrospect, it seems that Scheele was first but he failed to publicise his discovery, while Lavoisier was later than Priestley but named the gas oxygen, finally bringing an end to the phlogiston theory.

Priestley publicised his gases research in two volumes of *Experiments and Observations on Air*, of which the latter volume, published in 1776, contained his discovery of dephlogisticated air. Three years later Priestley fell out with Lord Shelburne, partly over those outspoken views of his concerning British attitudes towards American colonists, with whom he expressed great sympathy. Shelburne was now charged with the onerous task of improving relations with the former colony, and Priestley's dissent must have been an embarrassment. Priestley moved his family to Birmingham where, closer to friends from his Daventry and Warrington days, he fell in with a group that called itself the Lunar Society and continued his research.

In the Lunar Society, Priestley at last found himself surrounded by like-minded men. This informal society of natural philosophers, intellectuals and prominent businessmen met for dinner every month on the Sunday closest to the full moon so that, in the absence of street lighting, the moonlight could ease the journey home.

There were many literary and philosophical societies in the provinces during the early years of the Industrial Revolution, but the informality of the Lunar Society of Birmingham set it apart, the conviction of its members that conversation was a fertile source of self-improvement drawing them together.

Cheerfully referring to themselves as lunaticks, its members included the manufacturer and engineer Matthew Boulton; an arms manufacturer called Samuel Galton; James Keir, an inventor chemist, geologist and industrialist; Charles Darwin's two grandfathers – the physician, inventor and philosopher Erasmus Darwin, and Josiah Wedgwood, famous for industrialising pottery production; and James Watt, fresh from working in Glasgow in the same university department as Joseph Black.

The members would argue long into the night about scientific conundrums, such as why thunder rumbles. Boulton tried to answer the question by filling a large varnished-paper balloon with a mixture of common air and Cavendish's inflammable air, then lighting a fuse beneath it and releasing it into the evening sky. When the expected bang failed to materialise, Boulton and his pals assumed the fuse had gone out and started talking among themselves. Their discussions were soon interrupted by a massive explosion, but in the excitement of it all they forgot to listen for the rumble.

Other experiments were much more successful and members of the group made many important discoveries and inventions, the spirit of their endeavours shown in Darwin's letter to Boulton, regretting that he was going to miss the next meeting: 'Lord, what invention, what wit, what rhetoric, metaphysical, mechanical, and pyrotechnical, will be on the wing, bandied like a shuttlecock from one to another of your troup of philosophers.'

Second only to the Royal Society as a gathering place for scientists, inventors and engineers in the late eighteenth century, the Lunar Society played a pivotal role in marrying the latest discoveries of experimental science to the new technologies of the Industrial Revolution. Supremely confident, its members believed that science could improve the lot of all humankind through increasing production and adding to material wealth.

Like Priestley, many of the members of this revolutionary committee of the Midlands Enlightenment were nonconformists who sympathised with the aims of slavery abolitionists, the French revolutionaries and the American anti-colonialist rebels. They were sincere in wishing to improve the lot of ordinary people. And although most of Priestley's colleagues' main interest was the application of scientific discoveries to industrial processes, they also indulged Priestley's investigation of science purely for science's sake.

In this environment, Priestley flourished, publishing several papers, but remaining stubbornly attached to the phlogiston theory, which by the mid-1780s had been comprehensively destroyed by Lavoisier, a frequent correspondent with the Lunar Society. While Lavoisier had established a new chemistry based around elements, compounds and a logical chemical nomenclature, Priestley and other members of the Lunar Society refused to diverge from their attachment to the discredited theory.

Priestley continued to publish research that he said supported the phlogiston model. His self-education in science and his religious views perhaps led him to search for experimental evidence to support his beliefs, rather than adhering to the proper scientific method of postulating a theory on the basis of existing evidence, then repeatedly modifying the theory through a process of experimental verification and testing to destruction. In some ways, Priestley was a relic of an earlier age in the history of science; he was a master experimentalist who made crucial discoveries, but his analysis of some of them was weak.

Simultaneous with his science publications, Priestley was busy during his time in Birmingham with theology, publishing a series of increasingly provocative books that were particularly influential on Rational Dissenters such as Thomas Jefferson, author of the Declaration of Independence and a future President of the United States. As a result, Priestley and his fellow Dissenters became increasingly regarded as radical republicans, particularly by Pitt the Younger's government, which was concerned by their support for the French Revolution.

When Priestley attended a celebration on 14 July 1791, held at a Birmingham hotel to mark the second anniversary of the fall of the Bastille, a rioting mob of political and commercial opponents attacked him and torched his house. His provocative response at his sermon the next Sunday was to quote Jesus on the cross – 'Father, forgive them for they know not what they do' – and then to flee with his family to London. But with the French Republican government declaring Priestley a French citizen as it went to war with Britain, the capital proved no safer for Priestley, who was burned in effigy for declaring that the French Revolution was a harbinger of the Second Coming of Christ.

In 1794, a week before Lavoisier was executed for treason in France, Priestley emigrated to America, where, until his death, he avoided political, religious and scientific involvement and devoted himself to writing about education.

*   *   *

The last of the great British pneumatic chemists of the eighteenth century, and the first to embrace Lavoisier's new chemistry, was a high-achieving Cornishman whose association with a miner's lamp, although laudable, did him a disservice in the history of British science by overshadowing his many other greater achievements.

Born in 1778, the eldest child of a woodcarver in Penzance, Humphry Davy became Britain's first fully professional scientist. But that was only the half of it. With an easy and entertaining manner, romantic good looks, a taste for hallucinogens, rumours of a stimulant addiction, and friendships with contemporary cultural figures such as William Wordsworth and Samuel Coleridge, Davy could better be described as the first bohemian scientist.

A published poet with a handsome, rugged appearance from his outdoors life in Cornwall, Davy was so good looking and charismatic that when he moved to London the city's most beautiful and beguiling women fought for space in his lectures at the Royal Institution, set up in 1799 as an organisation to facilitate scientific education and research, particularly to alleviate poverty. Founded by the leading scientists of the age, including Cavendish, the Royal Institution was initially financed by Sir Thomas Bernard, a leading philanthropist, and the Society for Bettering the Conditions and Improving the Comforts of the Poor. The popularity of Davy's lectures transformed the precarious finances of the fledgling establishment. Carriage traffic bringing female fans to his lectures at the Royal Institution's magnificent premises in Albemarle Street was so heavy that the street, directly off Piccadilly in the West End of London, had to be converted to one-way operation on those nights.

But Davy was much more than just a pretty face. In the course of a long and distinguished career, he discovered and investigated several gases, rewrote our understanding of acids, founded the world's first Geological Society, made valuable applications of chemistry to agriculture, invented electric light and saved the lives of countless miners with his safety lamp.

Like so many of his scientific peers and predecessors, Davy was shaped by a somewhat unconventional childhood. From a young age, he was fascinated by the outdoors, learning to fish and shoot before he reached school age, and he was also a natural storyteller with a remarkable memory.

By the age of eight, Davy was a regular fixture in the market-place of Penzance, where he would stand on a cart in front of groups of children, relating the content of the most recent books he'd read, in particular *Pilgrim's Progress*. The 'applause of my companions was my recompense for punishments incurred for being idle', Davy later wrote, but he turned this idleness to a virtue.

Understimulated at school, Davy became highly self-motivated and later in life attributed his talents to this period of self-discovery, of which he said, 'I consider it fortunate I was left much to myself as a child, and put upon no particular plan of study … What I am I made myself.'

When Davy was nine, his parents moved to a farm a few miles outside Penzance, so Humphry boarded with his godfather, an apothecary called John Tonkin, who lived in Penzance and soon sent him to a private grammar school at Truro. A turning point came in 1794, when Davy's father died, leaving his widow with debts ten times the size of her annual income. Humphry told his mother not to worry; he would provide for his four younger siblings. Within months he had taken an apprenticeship with a Penzance apothecary-surgeon, Bingham Borlase, and set about teaching himself anatomy, chemistry, botany, physics and mechanics, while acquiring a working understanding of French from a visiting priest.

In the winter of 1797, Davy read Lavoisier's *Traité élémentaire de chimie* in its original language. This textbook of chemistry made total sense of the idea that oxygen was the crucial compo-nent in air for combustion. But it also appeared to make modern chemistry an entirely French science. Inspired by Lavoisier, Davy was determined to reclaim chemistry for the English.

Early in the new year he returned home to find that his mother had taken in a lodger to help pay off his father's debts. The son of James Watt, the steam engine inventor (about whom, more later), the lodger was a consumptive young man called Gregory who had studied chemistry in Glasgow. Davy was initially suspicious of this youngest member of the Lunar Society in Birmingham, but soon overcame his scepticism to form a close friendship based as much around drinking as a mutual interest in the natural world. Together they explored the local slate and tin mines, often a risky enterprise in excavations that were unstable and that ran deep beneath the sea bed. Back at Watt's mother's home, they conducted a series of experiments investigating heat, aiming to disprove Lavoisier's theory that heat was a weightless chemical element.

With the assistance of James Watt and the local member of parliament, who had an interest in science, Davy sent a paper, which included his theory that heat was motion, to a friend of theirs, a chemist called Thomas Beddoes. Politically radical, Beddoes had studied under Joseph Black in Edinburgh and knew several members of the Lunar Society, but his republican views had ended his career at the University of Oxford. Immediately impressed by Davy's work, Beddoes offered him a position at the Pneumatic Institute, an establishment financed by Josiah Wedgwood that he was setting up in Bristol to investigate the therapeutic properties of gases.

Throughout this time, Davy had not abandoned his interest in literature, often reciting poetry as he walked through the wild Cornish countryside. Contemplating his next move, Davy's dreams and aspirations were reflected in his best-known poem, written two years earlier and published in 1799, in which he wrote of hoping '*To scan the laws of nature, to explore / The tranquil reign of mild Philosophy / Or on Newtonian wings to soar / Through the bright regions of the starry sky.*'

Davy was not quite 20 years old when he arrived in Bristol and he threw himself into his research with gusto. Beddoes believed inhaling gases such as hydrogen might cure or alleviate common ailments such as constipation, so Davy set about synthesising various gases and, in true experimentalist fashion, testing them on himself.

In 1800, he discovered nitrous oxide and, shortly thereafter, the giddying effects of inhaling what was then known as laughing gas. Deep-inhaling four quarts of nitrous oxide from a silk bag, he found it lowered inhibitions and led to wild swings of emotion. He wrote of the experience:

> My emotions were enthusiastic and sublime, and for a minute I walked about the room perfectly regardless of what was said to me. As I recovered my former state of mind, I felt an inclination to communicate the discoveries I had made during the experiment. I endeavoured to recall the ideas, they were feeble and indistinct; one collection of terms, however, presented itself: and with the most intense belief and prophetic manner, I exclaimed ... 'Nothing exists but thoughts! – the universe is composed of impressions, ideas, pleasures and pains!'

According to some reports, Davy eventually became hooked on nitrous oxide, clearly oblivious to its addictive properties when he stated that it had all the advantages of alcohol with none of its flaws. He was supposed to have taken the gas up to four times a day for most of the rest of his life.

Since arriving in Bristol, Davy had fallen in with a racier crowd than he was used to in Cornwall. Through Beddoes, he had established a close relationship with Samuel Coleridge, and a loose friendship with another poet, Robert Southey. With the two poets, Davy experimented with laughing gas, which was fast becoming the LSD of its day among hedonists, Southey writing of being 'turned on' by its effects. Sometimes Davy would drink an entire bottle of wine in one draught before inhaling the gas. The effect, he wrote, was a 'highly pleasurable thrilling ... the objects around me became dazzling and my hearing more acute'.

Recognising the pain-relieving effects of nitrous oxide, Davy also suggested 'it may probably be used with advantage during surgical operations', although dental patients would have to wait more than forty years before they could enjoy its charms. However, not all Davy's gas experiments were quite so enjoyable. Four quarts of hydrogen nearly suffocated him; inhaling nitric oxide he found extremely painful, most probably because it would have combined with oxygen and the vapour in his lungs to form nitric acid; and carbon monoxide, the car exhaust gas used in suicide attempts, would have killed him had he not dropped the gas-bag mouthpiece as he keeled over onto the floor.

Davy published the result of his experiments, including commentaries by Coleridge on the gases' effects, in 1800. With graphic details of the effects of nitrous oxide on Davy's consciousness, the book also described his experimentation with gases on other human subjects, including young women who visited the laboratory, as well as live birds, rabbits, cats and dogs. Critical reaction to the book was mixed, ranging from academic indifference to scandalous glee at suggestions that Davy, Coleridge and Southey had held orgies at the laboratory, using nitrous oxide to seduce their victims. Described as 'bladder conjurors and newfangled Doctors pimping for Caloric', their supposed antics even prompted a satirical poem, *The Pneumatic Revellers*, in which the revellers 'cried, everyone, 'twas a pleasure ecstatic, To drink deeper drafts of the mighty Pneumatic!'

In an attempt to resuscitate his reputation, Davy followed the book with a paper the following year in the Royal Society's *Philosophical Transactions* on the acid battery he had developed. It brought him to the attention of Count Rumford, an American-British physicist involved in setting up the Royal Institution. Rumford, on the look-out for a chemistry lecturer, was initially unimpressed by Davy, until he heard him speak and was convinced by his natural charisma. Davy was offered the position of assistant lecturer in chemistry, director of the chemical laboratory and assistant editor of the Royal Institution's journals.

Within a year of joining the Royal Institution, Davy was a professor and the sensation of smart London. Often he would dash from his laboratory, donning a clean shirt on top of his dirty one, then deliver a highly polished lecture on scientific and technological progress entirely from memory.

Initially, Davy worked on agricultural chemistry and ways of improving the tanning of leather, but he soon moved on to the area in which he made his international reputation: electrochemistry.

Contrary to the established thinking at the time, Davy thought electricity was a force and not a substance, and that the forces holding compounds together were essentially the same as electrical force. This meant that compounds could be split apart by applying electricity, a process called electrolysis that had already been proven to work on water.

Davy became a master pioneer of electrolysis, designing his own battery and using it to split common compounds, such as magnesia, potash, lime and soda. In the process, he isolated several previously undiscovered metals, such as potassium from potash, sodium from molten sodium hydroxide, calcium from lime, as well as magnesium, boron, barium, strontium and aluminium – or alumium as he called it at first, later updating it to the name adopted by Americans, aluminum, before it was changed to aluminium to match the -ium pattern of other metals he had isolated.

Davy also produced an electric arc, which twenty years later he adapted into a design for an arc light, the first time electricity had been used for lighting. For his electrolysis discoveries, he was awarded a Napoleonic prize from the Paris Académie des Sciences, quite an achievement considering Britain was at war with France.

Davy went on to discover chlorine in 1810. He then investigated hydrochloric acid, in the process disproving Lavoisier's contention that all acids contained oxygen. He insisted that acids all contained hydrogen instead, a hypothesis that turned out to be entirely correct.

However, in 1812 Davy nearly killed himself in the laboratory (albeit not for the first time) in an experiment on the first high explosive, nitrogen trichloride, recently isolated by Pierre Louis Dulong, who had sacrificed two fingers and an eye in the name of science.

Davy was luckier. He damaged only his eyesight, so he hired an assistant, a young bookbinder from a poor background who had approached him after one of his lectures. The bookbinder's name was Michael Faraday and Davy's employment of him was later regarded as probably his greatest contribution to science.

Later that year, Davy resigned from the Royal Institution. After all his self-inflicted experiments with gases, he was a wrecked man, a condition in which his laughing gas addiction most likely played a part. By then he had discovered twelve elements, a fifth of the known total. He was knighted for his efforts a few days before he married a wealthy widow and the next year went on a tour of Europe, during which he picked up his French medal. Davy was lauded wherever he visited, particularly as he often used the opportunity to make new discoveries. In Paris, he determined the properties of iodine. In Florence, he focused the Sun's rays to ignite a diamond, thereby proving it was pure carbon.

When Davy returned to Britain in 1815, the Society for Preventing Accidents in Coal Mines approached him. As Britain's leading chemist, he was asked if had a solution to the problem of explosions in mines. The huge demand for coal in the early nineteenth century had brought with it a terrible toll in lives, most of which were lost in underground explosions. The principal cause was firedamp, thought at the time to be hydrogen. Miners carried caged canaries with them to detect the gas, but it was often too late by the time the canary fell off its perch.

After a few weeks' research, Davy determined that firedamp was a mixture of air and methane that had seeped out of coal seams, and that it would explode only at high temperatures. His invention was a new form of miner's lamp fitted with a two-layer metal gauze chimney that surrounded the flame. Oxygen could still reach the flame, but by dissipating the heat of the flame, the metal chimney prevented firedamp from reaching its detonation temperature.

Davy's lamp was one of the first examples of a research scientist being asked to apply his skills to an industrial product. Sir Joseph Banks, President of the Royal Society, wrote a congratulatory letter to Davy, saying his lamp would place the Royal Society higher in popular opinion than all the arcane discoveries beyond the understanding of ordinary people.

Davy refused to patent his discovery or profit from a humanitarian product intended to reduce the risk of a perilous occupation, although that didn't stop George Stephenson from trying to patent a very similar design.

In 1818, Davy was made a baronet in recognition of his achievement. Two years later, on Banks's death, Davy was the obvious candidate to replace him as President of the Royal Society.

Davy's last years, during which he suffered two strokes, were marred by a falling out with his former protégé, Faraday, of which more in Chapter 10, but it prevented the younger man from fulfilling his full potential until after Davy's death. Despite his ill health, however, Davy was not idle. As well as investigating ways to prevent corrosion of warship hulls, he returned to his roots by writing a book on natural history, but he was worn out. He died in 1829, aged 50.

Chapter 7

POWER FROM PROGRESS

Davy's lamp was certainly one of the earliest examples of a British research scientist being commissioned to solve an industrial or social problem, but it was far from the first time that a British scientific discovery led to the creation of a major British invention. More than 100 years before Davy created his miner's lamp, the inventors of the steam engine drew on the findings provided by Boyle and Hooke's air pump. Their explanations of the atmosphere and the nature of vacuums informed the designs of the first generation of steam engines, while James Watt used Joseph Black's work on latent heat to refine the steam engine into a machine that could do genuinely useful work.

Although Watt is most usually associated with the invention of the steam engine, he does not in fact deserve the credit. What he did was to design the first truly efficient engine to make proper use of the power of steam, but – as with many inventions – the steam engine developed incrementally, each stage incorporating new scientific discoveries. In fact, the first steam engine patent was granted in 1698, the year that James Watt's father was born, but the story begins earlier even than that.

By the early sixteenth century, Britain had cut down most of its forests so that Henry VIII and the Royal Navy could build the fleet that defeated the Spanish Armada. Wooden housing and charcoal for firing blast furnaces added to the wood shortage, so that by the mid-sixteenth century the country's timber stocks were so depleted that Parliament passed laws restricting the use of wood. Britain was importing timber from Scandinavia, quite an irony for a nation built on huge reserves of coal. However, the extraction of coal required deep mines, and the curse of mines was their tendency to fill with water.

The need to find a way of emptying mines of water coincided with the Newtonian revolution in science and the flowering of quantitative science that followed in Newton's wake. Boyle's work on gases led the French physicist Denis Papin to develop a steam digester, the forerunner of the pressure cooker, which he demonstrated to Charles II and the Royal Society by preparing a pressure-cooked meal. Papin realised that the steam pressing against the lid of his cooker would be capable of driving a piston, and although his designs for a steam-driven machine were not practical, they inspired others to develop steam engines.

Thomas Savery, an engineer from Devon, realised that he could build upon Papin's idea if he also exploited Hooke's discovery that air pressure could be put to work if a vacuum was created. Called the Miner's Friend, Savery's primitive engine had a large cylinder in which water was heated until it formed steam. A valve on the cylinder was closed and cold water was poured over the cylinder to force the steam to condense, which created a vacuum. When the valve opened, the vacuum 'sucked' water up from the mine into the cylinder (although technically, the imbalance between the pressure of the atmosphere and the pressure within the cylinder caused atmospheric pressure to push the water up the pipe). The cylinder would then be heated and the cycle would start again.

Strictly speaking, Savery's machine was an atmospheric engine, not a steam engine, but it was nevertheless the first machine to make use of steam under pressure. Savery built several prototypes, one of which he demonstrated at the Royal Society, but there is no evidence of his Miner's Friend being installed at a mine, most probably because the technology at the time was not capable of building vessels that could contain high-pressure steam reliably for a long time.

* * *

Within fourteen years of Savery's patent 'to raise water by the force of fire', a blacksmith from Dartmouth had built upon his fellow Devonian's idea to create the first proper steam engine. As a dealer in iron tools to Cornish tin mines, Thomas Newcomen had seen first hand that the established practice of using horses to lift buckets of water from the mines was hopelessly inefficient, and he realised that something better was needed. By 1712, he had the answer.

Like Savery's Miner's Friend, Newcomen's machine was an atmospheric engine that engaged the weight of the atmosphere to produce the working action. From afar it looked like one of those oil pumps that resembles the nodding head of a donkey; it had a long beam on a pivot that seesawed slowly up and down. At one end of the beam, a long iron rod descended deep into the mine, where it connected with a pump at the bottom of the shaft. The other end of the beam was connected to the shaft of a piston that was sunk into a large cast-iron cylinder.

The dimensions were vast: many of the cylinders in subsequent machines built by Newcomen were as much as 12ft high and 6ft across, such a vast size being needed to generate enough power in what was a very inefficient engine. A large boiler fed steam into the cylinder, thereby pushing the piston up, aided by a weight at the other end of the pivoted beam. When the cylinder was full, the steam supply shut off and cold water squirted into the cylinder. The steam condensed, generating a partial vacuum. In spite of the counterweight, the pressure of the atmosphere was sufficient to drive the piston down, thereby lifting the pump rods to suck water out of the mine. When the piston reached the bottom of the cylinder, a valve opened, letting steam back into the cylinder and the cycle started again. It wasn't fast and it wasn't pretty, but it worked. Unlike Savery's machine, the working stroke of the engine could be repeated automatically and indefinitely, achieving maybe five or six strokes a minute, day in day out, as long as it was fed steam.

Although his ironmongery and smithy were in Dartmouth, Thomas Newcomen installed his first working engine at a coal mine at Coneygree coalworks at Dudley Castle, near Wolverhampton. According to some accounts, Newcomen and his assistant, a plumber called John Calley, stumbled by accident upon a more efficient method of condensation. Having noticed that one of Newcomen's engines suddenly increased in speed without the usual supply of condensing water, they stopped the machine and investigated. Newcomen and Calley discovered a small hole in the piston. Through it, water was entering the piston and instantly condensing the steam. The two men immediately modified the engine design so that an overhead tank injected a small jet of water directly into the inside of the cylinder. This improvement greatly increased the efficiency of steam engines and was soon adopted by many other manufacturers. A second innovation, an automation of the valves controlling the steam and water, was devised by a young boy, Humphrey Potter, who was employed to guard the engine. Certainly, neither Newcomen nor Calley had been educated in science or mathematics, so they had to use trial and error to determine the size of piston and cylinder necessary to power their pump.

Newcomen's monster machines soon caught on, particularly after Henry Beighton, a Warwickshire surveyor, calculated tables to show the size of cylinder required to raise a particular amount of water by a particular height. He also invented a safety valve for the engine, adapted from the safety valve on a pressure cooker and designed to prevent excessive pressure from causing the boiler to explode.

Local millwrights and blacksmiths assembled the rough-and-ready apparatus at many locations, some of which remained in operation well into the early twentieth century.

Newcomen's engine supplied the coal industry with the tool it needed to expand massively and supply the burgeoning demand for coal. Moreover, it used the small lumps of coal that were a waste material at collieries.

* * *

While Newcomen was developing his machine and touting it around mines, James Watt was working as a laboratory technician and instrument maker at Glasgow University, where Joseph Black was professor of chemistry.

Watt's career was in the doldrums. Born into a family in which five of his seven siblings died in infancy, he'd been a weak and shy child prone to migraines and considered slow at school. The only areas in which he showed any natural ability were mathematics and handicrafts. He was expected to join his father's chandlery and carpentry business, but his father's business faltered, so Watt found employment with a Glasgow optician and as a handyman. Then, at the age of 19, he travelled for twelve days on horseback to London to take up a demanding and poorly paid apprentice-ship as a maker of laboratory instruments.

Watt was a good pupil. He learned new skills and techniques quickly, but the impoverished physical conditions of his appren-ticeship soon affected his health. The navy press gangs who were forcibly recruiting non-Londoners posed an even greater poten-tial threat to his well-being, so he fled the capital. In 1756 he returned to his hometown of Greenock where, barred from setting up his own business because he had not completed his appren-ticeship, he eventually found a job at the university servicing labo-ratory equipment. Eight years later, the university's natural philosophy department asked him to repair a model of a Newcomen engine. Fixing the machine was no trouble for Watt, but he was convinced he could improve on its hopeless inefficiency. Watt realised Newcomen's design had two problems: air entered the cylinder, reducing the effect of the partial vacuum, and it needed a lot of coal to maintain a sufficient head of steam.

Watt built his own model of a Newcomen engine and embarked on a series of experiments to investigate the properties of steam and the effect of changes in the dimensions of the cylinder, discussing the engine with Joseph Black. He thought back to when, as a child, he would sit at his aunt's tea table, watching a kettle boil on the stove. Why was it, he'd wondered as he'd caught droplets of water from its spout, that steam so quickly condensed back to water? Although he hadn't known it at the time, he'd been watching the latent heat of water in action. The problem with Newcomen's engine, Watt now realised, was the way in which the steam was cooled to create the partial vacuum. Spraying water into the cylinder to condense the steam also cooled the cylinder. To reheat the steam, he had to raise the temperature of the whole cylinder above the boiling point of water, which required a lot of energy, much of which was being expended in latent energy.

In May 1765, six months after he'd started investigating steam power, the solution occurred to Watt during a Sunday afternoon walk on Glasgow Green. 'I had not walked further than the Golf-house when the whole thing was arranged in my mind,' he wrote. 'All improvements followed as corollaries in quick succession, so that in the course of one or two days the invention was thus far complete in my mind, and I immediately set about an experiment to verify it practically.'

As James Watt strolled across Glasgow Green worrying about how to overcome the wasted energy in Newcomen's steam engine, the answer suddenly came to him. Why not put the condenser separate to the main cylinder and piston?

It meant he wouldn't have to heat and cool, heat and cool the cylinder and piston with every stroke. It meant the engine would be more efficient – three times more, it turned out – than the Newcomen engine.

Fired with enthusiasm, Watt rushed back into his workshop to build his separate condenser. He used a brass syringe for the cylinder. The condenser he made out of two pieces of tin and a thimble. And in a few days he'd got the system working. But although it was a moment of genius, it would be another ten years before he had a full-size working steam engine.

James Dyson

Watt's great breakthrough was to exploit what he called the 'elastic quality' of steam and divert it into a separate condenser. With lagging, Watt could keep the cylinder hot and the condenser cold to improve the thermal efficiency of the engine. And by using a separate condenser he could reduce the engine's energy requirements and running costs while increasing the speed of the engine.

Watt's development of his steam engine design was delayed by having to take on a new job as a surveyor for the Scottish canals, but by the autumn of 1768 he was ready to build his first prototype in a tiny cottage behind the large house of his financial backer, a doctor called John Roebuck. Roebuck had made his money in Birmingham with a sulphuric acid factory that he set up with a member of the Lunar Society. Interested in expanding into cast-iron manufacture, he had taken out leases on several coal mines and regarded a £1,000 investment in Watt's engine as a worthwhile means to improve output.

Although Watt could not get a steam-tight fit between the cast-iron piston and its cylinder, he took out a patent in 1769 for 'a new method of lessening the consumption of steam and fuel in fire-engines'. Two-thirds of the patent was assigned to Roebuck in return for an investment in Watt's development and legal costs. Watt needed further investment to improve his design, but Roebuck soon went bust and Watt was left in the lurch.

The previous year, Watt had stopped off in Birmingham during a trip to London and had met Matthew Boulton, a leading member of the Lunar Society. He'd explained his idea for a steam engine with a separate condenser and subsequently suggested to Roebuck that Boulton be sold a share in the patent in return for a financial investment. Now that Roebuck was bust, Watt suggested to him that he should sell his entire stake to Boulton.

Watt hated having to deal with the problems of financing further development of his engine, writing that 'nothing is more contrary to my disposition that hustling and bargaining with mankind, yet that is the life I now constantly lead', and his problems were far from over.

Roebuck refused to sell to Boulton, but by 1773 he had reneged on his commitment to increase his investment, thereby freeing Watt from his obligation. Watt dismantled the engine built behind Roebuck's cottage, transported it to Birmingham and the next year, following the death of his wife, moved to Birmingham himself.

Boulton was the best thing that could have happened to Watt. In return for two-thirds of any income from the engine, he financed all current and future development expenses and any manufacturing costs. He made a splendid array of craftsmen available to Watt at his Soho Manufactory, kept Watt focused on the steam engine and restrained him from going off in pursuit of other goals.

But still the problem of making the engine steam-tight remained. Then in 1775, Watt heard of a new boring machine invented by John 'Iron-mad' Wilkinson. Obsessed with making everything from iron, this larger-than-life character had built an iron chapel with iron window frames, an iron pulpit and iron pews, as well as an iron boat and even an iron coffin for himself. In the event, the coffin turned out to be too small when he finally needed to use it, and the undertakers had to blast a hole in the ground in order to bury its larger iron replacement. That lay in the future; in the present, Iron-mad Wilkinson had developed a new machine that could bore smooth cylindrical barrels into cannons. Highly impressed by the machine, Watt commissioned Wilkinson to make a cylinder for his steam engine. Totally steam-tight, it worked perfectly. Watt had his first working steam engine. For the next twenty years Wilkinson made all the cylinders for the Boulton & Watt steam engines, the blowers on his furnaces being driven by the second steam engine Watt ever made.

For the first few years, each engine was custom-built for each customer. While Boulton handled logistics and the business side of their partnership, Watt's role was to develop, design, draw and work on patent specifications. By the late 1770s, Watt had designed a genuine steam engine, sealing the cylinder at both ends to allow the expansive power of steam to replace the action of the atmosphere. Then he introduced steam both above and below the piston in the cylinder to make it into a double-acting engine in which the piston produced useful work with both strokes.

Watt created a whole new technology to make his engines. The future had arrived. Watt's monsters throbbed day and night and there seemed to be no limit to the power they gave to man.

In spite of all his improvements, Watt's engines were still beam engines, producing a simple reciprocating action ideal for pumping, but with few other applications. Urged on by Boulton, who wrote to him in 1781 saying that 'the people of London, Manchester and Birmingham are steam mill mad', Watt devised a method of turning the back-and-forth action of the piston into the rotary motion of a wheel. Equipping his engine with a set of sun-and-planet gears (a 'planet' cogwheel on the end of the engine's beam revolved around a larger 'sun' cogwheel that turned the drive shaft), he connected the beam to a spinning fly-wheel. Now the steam engine could power a wide variety of activities.

Watt's next refinement, a centrifugal steam governor to control the speed of the engine, became the first automation device, the modern term 'cybernetic' coming from the Greek for governor. With the governor and a parallel motion device that kept the piston upright in the cylinder, this engine was a huge success. It was rapidly adopted by industrialists, such as the cotton textile manufacturer Richard Arkwright, who until then had depended upon water or animal power to drive their heavy machines.

The ramifications for the British economy were huge. British industries no longer needed to situate themselves near a river to make use of the power of waterwheels. They could be positioned now wherever there was a reliable source of coal. And thanks to Newcomen and Watt, that source had become much more dependable.

The social changes brought about by steam power were just as dramatic as the economical and commercial ramifications. Cottage industries such as cloth weaving, sewing, lacemaking or small-scale manufacture, the mainstay of the economy for centuries, were made uneconomical by large-scale production in factories. Craftspeople working at home were replaced by factory workers; the countryside and farming withered while cities boomed and slums mushroomed.

Pragmatism and circumstance had provided the tipping point necessary to start the Industrial Revolution. Other factors, such as Britain's relative stability compared to its continental neighbours and its large population squeezed into a small, confined land mass, also played a part in making Britain the first country to industrialise. But once it started, the Industrial Revolution triggered a beneficial cycle between scientific discovery and the development of new technologies. Discoveries in heat, thermodynamics, mechanics and chemistry rapidly made their way into new inventions, which in turn made further discoveries possible.

Boulton and Watt produced about 500 engines altogether in the period of their partnership from 1775 to 1800. They transformed the machine from a simple steam-propelled pump into a sophisticated and versatile prime mover that could be applied to a very wide range of industrial processes, the first modern device to take energy from nature and apply it to almost any task.

When his patent ran out in 1800, Watt retired. Boulton and his sons had already taken over the running of the Soho Manufactory, so Watt returned to inventing new devices. Weighed down by administrative commitments during his partnership with Boulton, he had invented a chemical process for copying letters involving the use of a gelatinous ink for writing that could be pressed through to several additional sheets of paper. He used it extensively for letters and drawings. In retirement, he worked on a three-dimensional version that could sculpt irregularly shaped objects such as busts. The original and a block of stone were mounted on adjacent turntables, which were linked to rotate and tilt in harmony with one another. Meanwhile a probe connected to a drill was run over the original artefact, the drill moving synchronically with the probe to cut a facsimile. Although he drafted a patent for the machine in 1814, Watt had not filed it by the time he died in 1819.

Watt was the last survivor of the Lunar Society members. Increasingly haunted by the concern that his mental faculties might fail, he gladly took on projects that he considered worthwhile, refusing to accept payment for them (he also refused any honours, including a baronetcy). For the Glasgow Waterworks Company he devised a method of conveying filtered water across the River Clyde to the company's pumping station at Dalmarnock. And, aware of the effects that his engines had on the air and populations of Britain's rapidly industrialising cities, he patented a smoke-consuming furnace, going on in his retirement to investigate various health-improving devices. Given that tuberculosis had claimed the lives of his daughter Janet and son Gregory, his interest was perhaps unsurprising.

Watt was an obsessive. Like most men of genius, he had ferocious powers of concentration, which led some to see him as morose. But I prefer to think that he was never a satisfied man. He always saw how to improve something or he was always solving things.

As an inventor, I can completely understand how he felt. You're never happy. There's always this insuperable problem that's facing you, that you've got to overcome, and as soon as you've overcome that one, there's another one and another one. It probably turns you into a bit of a perfectionist.

The frustration that Watt must have felt about how badly things worked around him can be shown in some of the inventions he made.

It was very important to measure things accurately so he invented the first micrometer, a beautiful piece of equipment. He got irritated copying letters so he invented a letter-copying machine. He was frustrated that the steam engine would work faster and faster and not run at constant speed, so he invented the first centrifugal governor.

Ultimately, James Watt's invention changed the world. He provided almost unlimited power for mines, factories and transport – even the power to make electricity. He made our present.

James Dyson

As well as the legacy of one of the most successful and useful machines of all time – and the social and political changes that followed in its wake – Watt's name was immortalised in the unit of power. To help market his engines, Watt decided to compare their strength to the work that a horse could do. He found that a brewery horse could raise a 150lb weight by slightly more than three and a half feet in one second, which was equivalent to lifting one pound by 32,400 feet in one minute. He rounded the number to 33,000 and thereby defined the imperial unit of power, the horsepower, as 550 foot-pounds per second. When the metric SI unit system was adopted by most countries in the 1960s, the Renfrewshire inventor's name lived on in the modern equivalent of the horsepower – the universal unit of power, the electrical as well as mechanical watt.

* * *

Often overlooked in Watt's life story is the fact that some of his most notable improvements to his steam engine were in fact developed by a fellow Scot by the name of William Murdock. Industrious, imaginative and extremely modest, it was only Murdock's loyalty to the company of Boulton & Watt that stopped him taking his skills and insight elsewhere or exploiting his ideas for his own gain. Without Murdock, who was eventually rewarded with a partnership in the company, Watt's engines would not have been quite as successful.

After a solid but unremarkable education, Murdock arrived at Boulton & Watt in Birmingham in 1777 and immediately became a member of the Lunar Society. As principal pattern maker and assistant engine erector, in March 1779 Murdock was sent to Wanlockhead, the highest village in Scotland, a place inhabited only because lead, zinc, copper, silver and some of the world's purest gold at 22.8 carats, used to make the Scottish Crown, were mined nearby.

While building a Boulton & Watt engine for the Straitsteps Mine at Wanlockhead, Murdock made some changes of his own design, initially to the great irritation of Watt, who despaired at Murdock's 'disposition to amend'. Later that year, Watt sent Murdock from the Soho Manufactory in Birmingham to Cornwall to supervise the installation of his engines in various tin mines in the Redruth area.

Murdock spent eleven years in Cornwall, where he made several more changes to the Boulton & Watt engines, including inventing the crucial sun-and-planet gear that turned the back-and-forth action of the piston into the rotary motion of a flywheel. This invention turned what had been little more than an efficient steam-driven pump into a universal source of power for almost any type of machine.

Murdock also developed a D-slide valve that became the primary valve mechanism on steam engines and locomotives for the next century. Again Watt was sceptical about its benefits until after he retired, when he conceded that although 'he set his face against it … now I am satisfied it is an improvement after all as … it has rendered the machine much simpler and there are so fewer parts to go wrong'.

It was also Murdock's idea to turn the cylinder on its side and disconnect it from a beam to make the engine much lighter, more compact and more versatile. This design was adopted by the marine industry and used extensively in steam ships, including Isambard Kingdom Brunel's SS *Great Western*, the first steamship to cross the Atlantic.

Although Watt appreciated Murdock's technical expertise and the ways in which he had improved the efficiency of his products, writing that Murdock was 'the most active man and best engine erector I ever saw', he still despaired at Murdock's interference. The company made its money from speedy installations, not better-performing engines. So, when Murdock came to him with a working prototype of a self-propelled steam carriage that carried a fire shovel, poker and tongs around a room, Watt showed his annoyance in a letter to Boulton: 'I wish William could be brought to do as we do, to mind the business in hand, and let such as Symington and Sadler throw away their time and money, hunting shadows.'

Frustrated by Watt's stubborn refusals, Murdock was tempted to break away from Boulton & Watt to file his own patent on the steam carriage. But Boulton persuaded Murdock to stay with the company and return to Cornwall, where his Redruth neighbour, Richard Trevithick, went on to develop the first steam locomotive.

Back in his Redruth home, Murdock started to wonder about the coal and other fuels used to heat steam in the Boulton & Watt engines. Curious to discover to what other purposes they might be put, he heated wood, peat, coal and other fuels to see if he could drive off any gas. When he put some coal in a kettle, heated it and put a flame to the spout, he found the gas ignited.

Although Murdock immediately recognised the potential convenience of his coal gas – it could be transported by pipeline and controlled by adjusting the rate of flow – his ideas were ridiculed by those he told about it, including Watt.

Undeterred, Murdock persisted. He built a coal gas retort in his garden to light the interior of his house, which became the first building anywhere to be lit by gaslight. After returning to Birmingham in 1800, he lit the outside of the Soho works in 1802 and most of its interior the next year, although the coal gas produced in Birmingham needed further purification to avoid an unpleasant smell.

Murdock's invention caught on. He sold it to a cotton manufacturer in Manchester in 1804. By 1807 some London streets were lit by gas lighting, Murdock proudly writing to the retirees Watt and Boulton that 'fifty lamps of the different kinds' were lit and with 'no Soho stink'.

Although he won the Royal Society's Rumford Gold Medal for his discovery and exploitation of coal gas, he failed to capitalise on his invention, mainly because Boulton and Watt's sons decided not to market it. By-products of Murdock's process were subsequently exploited to make coke, ammonia, phenol, bakelite, the first synthetic dyes and aspirin. Meanwhile, Murdock continued working for the company, inventing a compressed air gun and compressed air power-tools, and investigating the exploitation of tidal power and even pedestrian power, but remaining relatively unrewarded for his endeavours.

* * *

Meanwhile, it fell upon Murdock's former neighbour in Redruth, Richard Trevithick, to exploit the idea of a steam locomotive.

In almost every way, Trevithick was a Cornish carbon copy of Murdock. He had a basic but good education and an aptitude for practical matters. Like Murdock, he was a powerful and large man, known widely as the Cornish Giant and apparently able to catapult a sledgehammer from the ground over the roof of a mine engine-house. And in common with Murdock, he was extremely inventive but failed to profit fully from his ideas.

Trevithick was deeply involved in the Cornish mining industry. His father was a copper miner, his brother was captain of several mines, and Richard married into one of the most significant mining and engineering families in the county.

In the early part of his career, Trevithick made several changes to improve the performance and efficiency of steam engines already in use at Cornish mines, which brought him into conflict with Watt over patent arguments. These tussles earned him the nickname Cap'n Dick from locals impressed by his insouciance towards one of the most powerful figures in eighteenth-century Britain.

When Watt's patents ran out in 1800, Trevithick went to town, developing a high-pressure version of Watt's double-acting engine, which he called a puffer because of its greater noise. For Cornish mines, which had no ready access to coal, the greater efficiency of Trevithick's puffers more than offset their greater cost. He then went on to design a 'Cornish boiler' as an upgrade for existing Boulton & Watt engines, and then the 'Cornish engine' in which both expansion and condensation took place using high-pressure steam. The Cornish engine was successful around the world.

At the same time as developing stationary steam engines, Trevithick was working on a high-pressure steam locomotive. Many Cornish mines already had tracks along which horses pulled wagons of ore, and some mines had already replaced wooden rails with iron ones. Trevithick spotted the obvious potential of steam-powered trains on tracks, but for the purposes of his demonstration engine he built a vehicle that could travel on the open road.

Using a relatively compact high-pressure engine, Trevithick built a steam locomotive at a foundry at Camborne. On Christmas Eve 1801, his steam-powered road vehicle had its first public demonstration, an event commemorated in the Cornish folk-song 'Camborne Hill':

> Goin' up Camborne 'ill, Comin' down,
> Goin' up Camborne 'ill, Comin' down,
> The 'osses stood still, the wheels went aroun',
> Goin' up Camborne 'ill, Comin' down!

Nicknamed the *Puffing Devil* by the locals on account of its 'devil-ish noise', Britain's first motorised form of road transport was short lived. Three days later, the *Puffing Devil* broke down on a journey to Tehidy, about three miles from Camborne. While Trevithick and his crew waited in a nearby pub, the locomotive overheated and was severely damaged.

Nevertheless, with the help of his fellow Cornishman Humphry Davy, Trevithick secured a patent on his invention and in 1804 he demonstrated the potential of iron tracks and steam power by building a locomotive that easily hauled more than twenty tons along the Pen-y-Darren tramway near Merthyr, albeit at the leisurely speed of 2.4 m.p.h. Looking to publicise his invention more widely, in 1808 Trevithick built a locomotive called *Catch Me Who Can*, which ran on a circular track in London's Euston Square and for a time was as fashionable an attraction as the London Eye is today.

In spite of his obvious success there was something of the fail-ure about Trevithick. Over and over again he failed to convince others of the undoubted worth of his ideas, which included steam-boats, steam-powered river dredgers and threshing machines. Eventually he gave up on locomotives altogether, emigrated to Peru to install pumping engines for silver mines, lost all his wealth and was bought a ticket home by George Stephenson's son, Robert.

To add further humiliation to Trevithick, Stephenson's money came from establishing the first inter-city railway service, which ran between Stockport and Darlington. In Trevithick's absence, steam railways had taken off. Initially they moved only goods. But by the time Trevithick died in penury in 1833, they were also conveying passenger traffic.

By 1850, Britain was criss-crossed by 6,000 miles of railway, the railroad was feeding the westward movement of the American frontier, and the iron rail was binding together great empires in India and Russia. Trevithick's invention of the steam locomotive had as profound an impact on Britain and other industrialising nations as Watt's invention of the steam engine. It brought cheap long-distance travel to the masses. Just as significantly, it meant that workers could afford to live away from their workplaces in new suburbs and it liberated cities from dependence on local food supplies.

Chapter 8

THE ORIGINS OF LIFE

For much of the scientific revolution, the biological sciences played second fiddle to the physical sciences. Although Willis and Harvey had made ground-breaking discoveries in human and animal anatomy, and Hooke's *Micrographia* had opened eyes to the extraordinary detail of the microscopic world, the science of biology lacked a grand universal theory to rival Newton's laws in physics. By the mid-nineteenth century, that was ready to change.

Drawing on two centuries of work by his predecessors, Darwin was ready to publicise a theory that would have as profound an impact on science and culture in general as Newton's law of universal gravitation had had nearly 200 years earlier. That such a large time span separated these two giants of science and the publication of their masterpieces is an indication of just how far biological sciences lagged behind the physical sciences at that time. And yet there is a direct link between the two men and their discoveries.

The story goes back to the triumvirate of the scientific revolution in Britain – Newton, Hooke and Halley – and May 1686, a month after Newton had delivered the first volume of his *Principia Mathematica* to the Royal Society. This was the moment at which Hooke raised the cry of plagiarism, the Royal Society backed out of its commitment to publish Newton's masterpiece and Halley stepped in to finance and personally oversee publication.

At the time, the Royal Society blamed economic reasons for not honouring its promise to Newton. In part, Samuel Pepys, President of the Royal Society, wanted to avoid involving the Society in the row between Hooke and Newton, but it was equally true that the organisation's finances at that time were highly stretched. This was because it had financed the publication of a beautifully illustrated book by one of its founding members, Francis Willughby, on the biological classification of fishes. Although this modest tome hampered the publication of the greatest ever book in science, *History of Fishes* was a significant book itself for several quite unusual reasons: it wasn't actually written by its ostensible author; it appeared long after the author's death; it sold hardly any copies (Halley was given fifty free copies in lieu of payment from the Royal Society); and, despite the above, it became a vital part of the jigsaw of discoveries and developments in biology that led directly to Darwin's theory of evolution.

In 1653, Francis Willughby was a new arrival at Trinity College, Cambridge. A competent mathematician, Willughby caught the eye of Isaac Barrow, who ten years later would befriend Newton and precede him as Lucasian Professor of Mathematics. But Willughby, who came from landed gentry, was more interested in the natural world than in mathematics. In 1655, he met John Ray, a lecturer at Trinity who shared his interest, and they became life-long friends.

Ray, the son of an Essex blacksmith, was an exceptionally bright man who, after teaching at Trinity for five years, briefly fell ill in 1650. While recuperating, he took long walks in the Cambridge countryside that cultivated an interest in the local plants and animals. By the time he returned to college later that year, Ray had decided to compile a catalogue of local plant types and species and to switch his academic pursuits from mathematics, the humanities and Greek to natural history.

In a garden at Trinity, Ray investigated the differences between species of plants and trees that he collected on long journeys around the British mainland and the Isle of Man. Travelling with Willughby and another former student, Philip Skippon, he also catalogued fish, birds, local customs, legends and antiquities. In 1660, Ray published *Catalogus plantarum circa Cantabrigiam nascentium* (*Catalogue of Plants around Cambridge*). With details of 558 plants and numerous local crops, it was the first attempt to list and catalogue species of plant with brief descriptions of their appearances, locations and uses.

Ray wanted to continue his work, but the Restoration intervened. As a non-conformist, Ray was forced to leave the university in 1662. Willughby immediately came to his aid, offering to
underwrite Ray's research and publications. The next year they
embarked on a three-year tour of Europe, Willughby studying
animal life including fish, insects, birds and mammals, while Ray
went plant hunting. Travelling as far north as the Baltic coast of
Germany and as far south and east as Malta, they catalogued and
collected thousands of specimens on a journey that has been
likened for its insights into the diversity of species to Darwin's
voyage on the *Beagle*. When they returned home in 1666, Ray
moved into Willughby's ancestral home, Middleton Hall at
Tamworth, staying there whenever he was not making further
extensive species-hunting trips around Britain. By 1670, he was
ready to publish *Catalogus plantarum Angliae* (*Catalogue of English
Plants*). With Willughby, he then set about collating a definitive
catalogue of every plant and animal known to them. But as
Willughby was preparing to leave on a species-hunting trip to
America, he fell ill. By July 1672 he was dead.

Devastated by the loss of his friend, Ray decided to devote the
rest of his life to producing a catalogue of life. Aided by an annu-
ity of £60 left to him by Willughby, he pressed on. In 1676, he
published *The Ornithology of F Willughby*, initially in Latin and
two years later in English. Although it was all Ray's work, he
published under Willughby's name as a tribute to his friend, a
practice he continued with *Historia piscium* (*History of Fish*).
Published by the Royal Society in 1685 at a cost of £400, it was the
book that Pepys blamed for depleting the Royal Society's funds
to such an extent that they could not afford to publish Newton's
Principia Mathematica.

Between the two Willughby-titled books, Ray published an
essay, *Methodus plantarum nova*, in which he set out the first clear-
cut classification system. It abandoned the previous system that
relied on the localities or medicinal properties of plants and
replaced it with a system based on physiology, anatomy and
morphology.

Ray's system ignored accidental or incidental features of plants
and concentrated only on the features that were essential to their
existence and always found in a particular species. His critical

division was between plants with a single seed leaf, such as grasses, lilies or palms, which are called monocotyledons; those with two seed leaves, called dicotyledons and comprising most other plants from daisies to oak trees; and flowerless plants called cryptogams, such as fungi, mosses and lichen. He then subdivided these divisions into families and defined the species as the ultimate fundamental unit of classification.

The apogee of Ray's classification system arrived in 1686, when he published the first part of his three-volume work, *Historia generalis plantarum*. With the subsequent volumes, published in 1688 and 1704, it described 18,600 mostly European species. It destroyed the classical explanation for the differentiation between creatures, which had been defined by Pliny, the Roman scholar.

Ray's work remained the definitive text on plant species until the time of Darwin, who was heavily influenced by it. It was rivalled only by the work of the Swedish botanist Carolus Linnaeus, who in the eighteenth century established the modern binomial system of classifying all species with a generic name and a specific name.

Ray followed the principles he had used for plants on his later classifications of animals, concentrating mainly on hooves, teeth and toes to delineate species of quadrupeds. Two books on classification of insects and of birds and fish were published posthumously.

* * *

Ray remained Britain's foremost botanist and species hunter until the eighteenth century, by which time plant collecting had become almost like a sport. With the British Empire at its height, we Britons had become an absurdly acquisitive nation, matched in our mania for discovering new species by Europeans and Americans. At a time in which international reputations were made by becoming the first to return home with a hitherto unknown new plant or animal, the most fanatical species hunters would set off to cover whole continents on foot, harvesting specimens along the way.

In the midst of this compulsion, one Briton emerged as one of the greatest explorers of the age and the man who put zoology and botany firmly on the scientific map. Born into a landed fortune, Joseph Banks differed from the undiscriminating gentlemen collectors of his day in that he was scientifically trained as a naturalist in the Linnaean mould – in other words, he had been educated in the taxonomy established by Linnaeus earlier in the eighteenth century.

As befitted a young man born into a large fortune, Banks was educated first at Harrow, then at Eton, where his lifelong fascination for botany first emerged.

In Britain, the amazing Sir Joseph Banks is hardly known, but in Australia and New Zealand he is something of a national hero and that's because he was with Cook when they were the first Europeans to land at Botany Bay. Banks was the man who recommended to the British government that they should send a few English convicts there to found a settlement. That settlement was to become the great city of Sydney.

Banks's path to becoming a great scientist began as a boy. He was already a keen collector of insects and birds' eggs, but the story goes that he found his true vocation one afternoon when he was 15 and he went for a swim in the Thames near Eton. And as he got out he was suddenly overwhelmed by the beauty of the wildflowers on the bank. He determined there and then he would become a botanist.

Banks started by paying the gypsy women who collected plants for the local druggists to teach him the names of the plants.

A few years later, Banks was inoculated with smallpox in the primitive and somewhat dangerous means of immunisation against the disease before the discovery of vaccination. It made him too ill to go back to school – but it didn't diminish his passion for botany.

David Attenborough

After school, Banks went up to Oxford, where he found the curriculum heavily biased towards the classics, so he dipped into his considerable fortune to finance a lectureship in botany. Although Oxford had a botany department, its lecturer, Professor Sibthorp, was reluctant to teach, so Banks leapt on his horse and rode to Cambridge, where he had heard there was a good botany lecturer called Israel Lyons.

Banks persuaded Lyons to move to Oxford. By employing a botanist from Cambridge University as his personal tutor, Banks had introduced teaching of the subject to Oxford, so it might seem surprising that he left Oxford without a degree. It was, however, a common practice for young gentlemen of means in those days.

When he moved to London, Banks's wealth and social standing gained him entrance to many of the most august institutions and he was soon invited to become a Fellow of the Royal Society. At the British Museum he met Daniel Solander, the assistant librarian, who, having trained under Linnaeus, passed on the principles of the Swede's classification system.

Banks's father died in 1761, leaving his only heir free to do exactly as he pleased with his vast inherited fortune, which included more than 200 farms stretched over large parts of Lincolnshire and Yorkshire. In 1766, at the age of 23, Banks found the opportunity he had been waiting for: an expedition to Newfoundland and Labrador on a fishery-protection ship under the command of his Old Etonian friend and future lord of the Admiralty, Constantine Phipps.

Following good Linnaean principles, Banks collected plant, animal and mineral samples on the trip and made extensive studies of the local human inhabitants. These were later catalogued at the Natural History Museum by his friend, Solander.

With the Newfoundland expedition deemed a success, two years later Banks found himself at the front of the list of naturalists being considered for Britain's most ambitious scientific expedition to date. This was the grand voyage to the Southern Pacific organised by the Admiralty and partly financed by the Royal Society in order to undertake observations of the transit of Venus across the disc of the Sun, as proposed by Halley some fifty years earlier. Halley, who had died in 1742, had predicted that the observations would make it possible to calculate the distance between Earth and the Sun, a crucial missing part in the jigsaw of longitude calculation essential for accurate navigation and the extension of British sea power.

Of all the applicants, Banks had the healthiest finances and best connections to get himself on board James Cook's first voyage to the southern hemisphere. Banks's berth aboard the *Endeavour* was ensured by the donation of nearly £1 million in today's money to the expedition's costs and a bawdy and louche friendship with Lord Sandwich, the first Lord of the Admiralty, with whom he was frequently seen socialising at London gentlemen's clubs or fishing in the company of 'two or three ladies of pleasure', according to the philosopher David Hume.

They sailed from Plymouth on 25 August 1768, Banks clearly having no intention of travelling light. Squeezed into the ship with him were Dr Solander from the British Museum; a scientific secretary by the name of Herman Spöring; two illustrators, one of whom specialised in landscapes, the other in natural history; two servants; a pair of greyhounds; a library of natural history books; and a panoply of exotic devices for finding, trapping and examining specimens.

Less than 100 feet long, the *Endeavour* was already overcrowded. With a crew of eighty-five plus the expedition's official astronomers and their equipment, stores for eighteen months at sea, a goat and the ship's cat, the sailors were allocated only 14in of hammock space. Captain Cook's Great Cabin was about 12ft square. But in spite of great differences in their characters and backgrounds, the famously bluff Captain Cook tolerated Banks remarkably well considering that he was forced to share his quarters with Banks's entire entourage and their scientific equipment, as well as all the specimens of fish, animals, birds and plants they would collect along the way.

> An uncomfortable voyage? Yes. But it was also extremely dangerous. As Dr Johnson put it, a ship was like a jail with the added risk of being drowned.
>
> Apart from shipwreck, there was also the danger of scurvy and other diseases, and fatal encounters with natives in unknown savage lands. Indeed, of Banks's party of eight, only three would return alive.
>
> When Banks left Plymouth in August 1768, it would be three years until he saw England again. He would come back a changed man, but in turn he would change the face of British science forever.
>
> David Attenborough

Surviving on a diet of pickled cabbage, albatross and other seabirds shot out of the sky by Banks, they sailed for Tahiti, Cook steadfastly ignoring Banks's repeated pleas to pull ashore in order to investigate the local wildlife. However, when in early January 1769 the *Endeavour* reached Tierra del Fuego at the southern tip of South America, Captain Cook conceded the currents around the Horn were dangerously fast and made for landfall until conditions could improve. While Cook and his crew restocked the ship's supplies, Banks and Solander went ashore, Banks commenting that the weather was ideal – 'like a sunshiny day in May' – for exploration. Having assessed the lie of the land from a nearby summit, the next day Banks set out with Solander, Spöring, Alexander Buchan (one of his artists), four servants and two seamen. Also with them were Charles Green, the ship's astronomer, and William Monkhouse, its surgeon.

Almost immediately, the party ran into trouble. As they fought their way through a dense boggy forest, which from afar Banks had assessed as tundra, Buchan collapsed with an epileptic fit. After stopping to attend to Buchan and to build a fire, Banks and Solander pressed on. Distracted by the discovery of some alpine plants, at first they did not notice that thick clouds had covered the sun. Then, as snow began to fall and the temperature plummeted, they immediately turned back to rejoin the rest of the party.

By the time they reached the main party, Buchan had recovered, but the snowfall was now so thick that they could not see further than a few feet ahead. With only meagre provisions and no proper shelter, they had no choice but to attempt to return to the *Endeavour*. Poorly prepared for the cold, Solander and one of the servants, Tom Richmond, soon grew tired and lethargic, a sign that hypothermia was setting in. Then, as darkness fell, Solander and Richmond collapsed. Banks eventually roused Solander and, leaving two men to watch over Richmond, helped him to a clearing where the rest of the party had set up a makeshift camp around a fire.

Later that night, one of the two men who had been guarding Richmond returned to Banks's camp around midnight, drunk and exhausted. Banks led four of his party back to where Richmond now lay with the remaining servant, George Dorlton. They built a crude shelter, then returned to their own camp, leaving Banks's two greyhounds behind to keep the men warm. It wasn't enough. By the next morning, Richmond and Dorlton were dead. To add to the tragedy, the storm having passed, Banks now realised they had spent the night tantalisingly close to the *Endeavour* and might have saved their companions' lives if they had hauled them back to the ship.

By midday, the depleted party was back on the ship, ready to sail around the Horn and on to Tahiti, where they anchored on 13 April 1769. While the astronomers prepared for their observations of Venus from a camp of several tents, a bakery, a forgery and an observatory, Banks and his assistants roamed the island, collecting flora and fauna. Meanwhile, the illustrators spent their days observing and painting, their tasks made more difficult by the vast numbers of flies and the Tahitians' desire to acquire anything made of metal.

'We can scarce get any business done for them,' Banks wrote of the flies in his journal. 'They eat the painters' colours off the paper as fast as they can be laid on, and if a fish is to be drawn there is more trouble in keeping them off it than in the drawing itself.'

As for the illustrators, Banks soon lost Buchan to another epileptic fit.

The Tahitians' quest for metal items was equally problematic. When they couldn't steal metal, the Tahitian women offered sexual favours in return for as little as a single iron nail, Cook complaining that most of his crew had caught venereal disease because Tahitian 'women were so very liberal with their favours'.

Cook was concerned not just about the health of his men. He knew that two years earlier, the crew of another ship, the *Dolphin*, had prised so many nails out of her timbers to exchange for Tahitian affection that the ship nearly fell apart in her next storm at sea.

Banks was also no stranger to the local charms, taking up with one of the queen's servants 'with fire in her eyes', although he justified it as all part of his Linnaean classification of an unexplored culture. In this, Banks seems to have been quite sincere. As well as physical experimentation with the locals, he learnt the basics of their language, documented and participated in their customs (including being tattooed and learning to play the nose-flute), and recorded the first European sighting of people surfing on waves. After departing Tahiti, the *Endeavour* sailed south to chart New Zealand, where Banks regarded the Maoris as too likely to commit cannibalism to risk going ashore.

When they reached the south-east coast of Australia, Banks was less reticent, finding so many species of plant for his collection that he named their landing site Botany Bay. From this future site of Sydney, Cook sailed his ship north, landing in October 1770 at Batavia, at that time the capital of the colonial Dutch East Indies, now the Indonesian capital, Jakarta.

Four months earlier, the *Endeavour* had nearly gone aground on the Great Barrier Reef and it now needed extensive repairs. While Banks and Solander headed inland to explore the Javanese countryside, thirty of the crew who remained in Batavia died of Java fever, a disease as deadly as malaria or yellow fever. Among the dead were Spöring and Parkinson, the expedition's last remaining illustrator.

The following June, after three years at sea, the *Endeavour* and her depleted crew landed back in Britain. Of his party of seven, Banks had brought only himself and Solander home. But he also had on board more than 30,000 specimens, of which around 1,400 had never been seen before. Fêted as a hero and completely over-shadowing Cook, Banks was introduced to George III. It was the start of a long friendship that Banks assiduously cultivated.

Cook and Banks were national heroes – for surviving, yes, but also for the wonders and the stories they brought back with them. They had mapped New Zealand and claimed eastern Australia for the crown; they called it New South Wales.

Banks was painted by Joshua Reynolds as a romantic hero. Young, handsome, charming and full of traveller's tales, he was the man of the moment. Stories of erotic adventures in Tahiti only increased the admiration and wonder he inspired in London society. Banks and his friends had brought back with them a vast haul of strange plants, mammals, birds, reptiles, insects, sea creatures and ethnological spec-imens the like of which had never been seen in Britain or Europe before. Many of those actual specimens have disappeared but there are lots of drawings and paintings to show us some of the wonders that Banks and his friends encountered.

Along with all the plants and flowers, birds and fish, Banks brought back the skin of a strange creature they had shot and eaten one morn-ing in Australia. He said that aboriginals called it kangaroo – and that's the name we use for it today. Banks commissioned the artist George Stubbs to stuff the skin and paint it. Stubbs had to make up a few details but the result was splendid!

The kangaroo caught the public imagination. And soon Banks would thrill London with another exotic creature. Banks had wanted to bring back a Tahitian with him, thinking in his own words 'to keep him as a curiosity, as my neighbours do lions and tigers'. And when Cook came back from his second voyage he did indeed bring a Tahitian, a young man called Omai.

Banks fitted out Omai with fashionable European clothes and took him to meet the king. Omai greeted George III with the words, "How do, King Tosh." And the king was delighted. He gave Omai a sword and recommended inoculation against smallpox, which Banks duly arranged.

Banks introduced Omai to everyone who mattered in London society. He went to the theatre and the opera, he dined with Dr Johnson, he played chess with fellows of the Royal Society. He even went ice-skating.

But after two years, fashionable society tired somewhat of Omai and when Cook set on his third voyage to the Pacific he took Omai with him as he had promised. However, when they got to Tahiti Omai was unable to readjust and sadly died two years later.

George III was fascinated by Banks and his discoveries. The two men became close friends – or at any rate as close as you can be to a king. In consequence, the king put Banks in charge of one of his favourite projects, the botanical gardens at Kew.

Now Banks sent out collectors in the name of George III and by the early 1800s Kew Gardens' reputation and influence had grown to such an extent that virtually no ship left India or any other British colony without some living or preserved specimen for the Botanic Gardens.

David Attenborough

With his many specimens, Banks was able to show that Australian mammals were quite different to the mammals on any other continent. They were nearly all marsupials and much less developed than the mammals with placentas found elsewhere. It was a profound discovery that would strongly influence the theories of Wallace and Darwin a century later.

Banks went on one more expedition to the North Atlantic and Iceland in 1772, this time with a mistress disguised as his valet, a common practice among seafarers at a time when women were not allowed to accompany ships to sea.

Through his close association with George III, in 1773 Banks was put in charge of the Royal Botanic Gardens in Kew. In 1778, he was elected President of the Royal Society. Three years later he was made a baronet. Now a statesman of science, he had a philanthropic attitude to the colonies. He took merino sheep from Spain to Britain and Australia. He introduced tea plants to India from China and had breadfruit brought to the West Indies from Tahiti, the ship carrying the fruit being the *Bounty*, whose crew famously mutinied against their captain, William Bligh. As well as being instrumental in developing Botany Bay as the first Australian colony, Banks supported many promising young scientists and turned his home in Soho Square into a salon for the capital's intellectuals.

By the turn of the nineteenth century, the descent of George III into madness left Banks without the support of his greatest champion and his influence waned. However, in having undertaken, supported, promoted and advised countless expeditions, he had established a tradition of scientific exploration that would lead to Darwin's voyage on the *Beagle* as well as many less notable trips.

In the Palm House at Kew stands the oldest pot plant in the world, *Encephalartos altensteinii*. It's only once borne a cone, in 1819, and Joseph Banks went to Kew especially to see it.

It proved to be his last visit. He was crippled with gout and died a few months later. He had asked that he should be buried without any ceremony and that his grave should not be marked by a headstone.

In accordance with his wishes there is no headstone but Banks does have a monument and you can see it in the herbarium in Kew Gardens. It's a plant that carries Banks's name, Banksia.

Joseph Banks put botany and zoology firmly on the scientific map. He revealed how rich and biologically diverse our planet is. He excited the imagination of the king and of the public. He made science in Britain really matter.

David Attenborough

* * *

Banks didn't keep all of the specimens collected on his travels for himself. While at sea on the *Endeavour*, the crew caught a giant squid. Hauled aboard, it was killed, then chopped up for the pot. After months with very little fresh food, they wanted to make the most of such a rich protein feast. Banks, Cook and the crew cooked and ate the entire animal. The only part of it that remained was its large, inedible beak, which Banks added to his extensive collection now cluttering Cook's quarters. When he arrived back in London, he knew exactly what to do with this strange memento of his time at sea. Banks passed the beak on to John Hunter – one of the most distinguished scientists and surgeons of his day – who was amassing an extraordinary collection of 14,000 anatomical specimens, including dissected humans and animals, in central London.

Modern histories of medicine often refer to John Hunter as a Scottish surgeon renowned for his use of unorthodox methods. In fairness to Hunter, every eighteenth-century surgeon used methods that would now be considered highly unorthodox. At that time, surgeons were more butcher than scientist. Cut and cauterise with a red-hot iron were the basic techniques. Hospital beds had handcuffs to restrain agonised patients. Finesse was frowned upon and speed was of the essence simply because there were no anaesthetics. The more fortunate patients were given laudanum, a heady mix of alcohol and opium. For the others, it was a case of scream and bear it. A good surgeon could remove a leg in 15 seconds. If the patient didn't die of shock, there was always a severe risk of infection. Surgeons saw no reason to wash their hands or even their instruments, which were often encrusted with blood and pus.

The problem was ignorance. Nobody knew how the body worked; surgeons simply did what they had always done: cut and hope for the best. John Hunter made a name for himself by becoming the first surgeon to take it upon himself to change all that.

After several years of working on the family farm in Larnarkshire, at the age of 20 Hunter rode on horseback to London to join his brother, William, already a surgeon and obstetrician. It was 1748 and surgery was only just becoming a respectable profession. Since the sixteenth century, it had been conducted by barber-surgeons, many of whom were illiterate and without formal learning. While physicians trained at university in order to be able to consult and prescribe medicines, these barber-surgeons usually picked up their trade as an apprentice to a more experienced colleague. When qualified, they were expected to perform any operation from cutting hair to removing limbs. Some took up residence in castles or stately homes, where they would treat the rich and powerful, but most made their money on the battlefield, treating soldiers wounded in combat.

The barbers' historic association with surgery is still seen in a relic from this time: the red and white barber's shop pole, originally intended to reflect the blood and napkins used to clean up during bloodletting. The United Barber-Surgeons Company had a long history dating back to 1540, when it was established by Henry VIII, but in the years shortly before Hunter arrived in London surgery had started to establish itself as a separate profession. Procedures had become more reliable and the work of surgeons was raising their status. Meanwhile, barbers were subjected to increasing restrictions forbidding them from conducting surgical procedures other than pulling teeth and bloodletting.

Then in 1745, three years before John Hunter arrived in London, George II separated the two professions. The London College of Surgeons was established and surgeons were expected to be university educated.

With very little formal education at all, Hunter had little hope of becoming a surgeon. Fortunately, he also had little intention of following in his brother's footsteps. His father having died, he had moved to London to live with his brother and was hoping to find any kind of work. However, William, who had studied anatomy and surgery in Edinburgh, London and Paris, offered him a position as his assistant.

After having presented and published his research into female reproductive anatomy at the Royal Society, William Hunter had set up a successful surgery and midwifery practice. When George II split barbers and surgeons, William Hunter advertised an anatomy course so that 'Gentlemen may have the opportunity of learning the Art of Dissecting during the whole winter season in the same manner as in Paris'.

Paris manner was a very discreet way of describing the kind of anatomical work prospective students could expect. It meant instruction on human corpses, mostly obtained from so-called resurrectionists (more commonly known as bodysnatchers or sack 'em up men), who under the cover of night raided graveyards for freshly buried corpses that they sold on to surgeons and anatomists.

William Hunter's course was already a success when his brother John arrived in London, and William went on to have a distinguished career in medical research. One of his investigations involved dissecting the corpse of a full-term mother to discover how the maternal and foetal blood systems interacted through the placenta. He was also one of the first to recognise the importance and significance of the survival of the fittest in reproduction, and he pioneered several discoveries in forensic medicine.

By the 1760s, William Hunter was surgeon to leading members of the aristocracy, including the royal family. He was a wealthy man whose collection of coins was said to be second only to that of the King of France. With a sizeable collection of anatomical specimens (smaller, however, than the eventual collection of his brother) and a library of more than 10,000 books, he bought a property in Great Windmill Street, close to Piccadilly Circus in central London, to use as a museum and lecture theatre for his anatomy and surgery demonstrations.

The building is still there. With a blue plaque commemorating 'Dr William Hunter, Anatomist', it stands next to the neon-drenched façade of the Windmill Theatre, where there are now nightly displays of human anatomy for less academic purposes. In the 1930s, the Windmill Theatre was the first London venue to put on risqué shows of naked young women; now it is a run-of-the-mill lap-dancing club in the heart of Soho.

With his brother William the leading anatomist of his generation, one might have expected John Hunter to have had an easy passage to his eventual status as the best surgeon in the profession's early history. But when he arrived in London in 1748, John faced a long and arduous struggle before he could claim that title.

In personality, the two brothers could not have been more different, but they played well to their strengths. Urbane, articulate and socially at ease in the capital, William's good humour and confident manner had played a significant part in making his anatomy courses so successful and in easing his passage through the salons of London. By contrast, John came across like a Glasgow navvy. Poorly educated and with a thick accent, he was more at home in the boisterous coffee houses surrounding Covent Garden market, where he made contact with the resurrectionists who supplied William's anatomy college.

Although William had the education, John was the more naturally talented dissector, making several notable discoveries in the decade that he worked as his brother's assistant, including insights into the growth and structure of bone, and the structure and function of lymphatic vessels.

With refrigeration unavailable to anyone without a ready supply of vast ice blocks, the Hunter brothers' reliance on freshly excavated corpses ruled out teaching in the heat of summer. So, in the first summer recess of the Hunters' anatomy school, John found himself a pupillage with a Chelsea Army Hospital surgeon regarded as the finest operator in London. The next year, he learnt at the side of St Bartholomew's splendidly named Percivall Pott.

There's a Joshua Reynolds painting of John Hunter in the Royal College of Surgeons. With an intense glare on his face, you can see just how obsessive and intrigued this workaholic was by the wonderful objects in biology that so fascinated him. While Banks was away exploring the other side of the world, John Hunter was exploring a new world of his own – the inside of the body.

Hunter's determination to drag surgery out of the Middle Ages and put it on a scientific basis meant making accurate maps of the body, and discovering how the various parts interacted and functioned. And that meant anatomising animals and humans, normal and abnormal. Hunter did hundreds of dissections. Little by little he built up a detailed knowledge of human anatomy, and upon this knowledge he based his surgical technique.

He soon built a reputation as a surgeon *less* likely to kill you than most of the others. He knew what he was doing and that's why he was willing to take on the cases nobody else would touch.

Robert Winston

In 1754, Hunter enrolled as a surgical pupil at St George's Hospital, at that time on Hyde Park Corner. Within two years he was a house surgeon. Although appointed a master of anatomy by the Surgeons' Corporation, he was still without the necessary credentials for membership of the Company of Surgeons, needed if he wanted to practise independently in London. In an attempt to gain the qualifications he had gone to Oxford, but had dropped out after only a few months because he was unwilling or unable to study the classics, complaining that 'they wanted to make an old woman of me, or that I should stuff Latin and Greek'.

Frustrated and facing a lifetime of intermittent employment, in 1760 Hunter joined the army as a staff surgeon. Britain was in the midst of the Seven Years War with France and desperately needed medics. For those prepared to enlist for a meagre salary, it promised eventual professional qualification as a surgeon.

When he arrived on the battlefield in Portugal, Hunter was horrified by the primitive state of the art. His patients were men torn apart by gunshot and shrapnel wounds and he could do very little for them. In some instances, doing nothing in the hope that nature might eventually heal the injury – a strategy frequently advocated by Percivall Pott – seemed better than even greater mutilation by the surgeon's knife.

Hunter's experiences in the army made him realise that an anatomist's understanding of the layout and appearance of organs was not enough in itself to perform effective surgery. He needed to understand *how* and *why* the various parts of the body worked. He needed to apply sound scientific investigation techniques to solving the mystery of the processes of life.

Hunter returned to London in 1763 with a determination to decipher human physiology. To support himself, he set up a surgical practice at Golden Square in the heart of Soho and drew a small salary from the army as a reservist, which eventually led to his appointment as surgeon-general of the army and surgeon-extraordinary to the king. He also continued to lecture on anatomy.

Ever since he'd started working as his brother's assistant, Hunter had retained animal and human samples from his anatomy investigations and classes. During his wartime surgical training he had continued taking specimens. Now needing more space to house his growing collection, he bought a small rural estate at Earl's Court, located at that time in the countryside at the edge of London. With first refusal rights to any dying animals in the king's menagerie, he soon built up a large array of live and preserved specimens in his garden, including a bull and a whale carcass.

Working obsessively, Hunter began his day at five o'clock, when he would conduct investigative dissections until nine. After breakfast, he would see his surgical patients until four o'clock. He would then return home for dinner and a brief nap before returning to his anatomy laboratory or study until midnight, often scribbling his notes on the backs of envelopes which he would later use as the basis of lengthy manuscripts dictated to his assistant.

Starting with the head, Hunter painstakingly worked out the purposes and functioning of the human body through comparative dissections of his human and animal samples. Working systematically, he moved on from the teeth to the mouth, throat, neck, chest, stomach and abdomen, analysing the organs, the circulatory, nervous and lymphatic systems, and the muscles and skeleton. To his detractors, who said he was obsessed with details, he responded with the maxim that from these details facts could be established, which in turn would explain the principles of the body's functioning and the causes of diseases. 'Without this knowledge a man cannot be a surgeon,' he said.

However, Hunter's most powerful retort was the array of surgical techniques he pioneered as a result of his anatomical investigations. His dissection of a deer's antlers provided an insight into collateral blood circulation that he later applied successfully to a female patient with an aneurysm behind her knee.

By 1770, Hunter's surgical practice had become sufficiently successful for him to take over his brother's property in Jermyn Street when William moved his school to Great Windmill Street. The building was large enough for Hunter to live there, to move his collection from Earl's Court, and to establish a surgery college at which Edward Jenner was one of his students. Scientists travelled from all over Europe to see his unparalleled collection.

Although he was now highly educated and respected among his peers, Hunter retained his famously gruff manner, impatiently advocating experimental investigation over logical deduction to a generation of surgeons, an ideology that had its roots in the scientific philosophies of Roger Bacon, William Gilbert and Isaac Newton.

'I *think* your solution is just,' he wrote in response to one of Jenner's requests for a comment on one of his theories. 'But why think? Why not try the experiment? Repeat the experiments. They will give you the solution.'

As a by-product of his obsessive collecting, Hunter refined William Harvey's technique of embalming by arterial injection, which led to him acquiring one of the more macabre and unusual specimens in his collection.

In 1775, Martin van Butchell, an eccentric dentist and former pupil of Hunter, turned up on the great surgeon's doorstep with a request that he embalm his recently deceased wife, Mary. His motivation was entirely selfish: under the terms of Mary's will, her husband could retain her considerable wealth for as long as her body remained above the ground.

Hunter and an assistant injected Mary with a mixture of oil of turpentine, camphorated spirit of wine and dyes to give the blood vessels of the face a lifelike colour. Glass eyes and a fine lace gown completed the procedure. Mary's corpse was then embedded in a thin layer of plaster in a coffin with a removable glass lid.

Butchell displayed his first wife's embalmed body in the window of his home and dental practice for several years. But on his second marriage he donated it to the College of Surgeons, where it remained, slowly deteriorating until it was destroyed in an air raid during the London Blitz in May 1941.

One story about Hunter has always intrigued me. It's about his interest in abnormalities and amputations. And intriguingly, it's hinted at in a portrait of Hunter by Reynolds.

Behind Hunter in the painting we can see a pair of feet in a glass case. These once belonged to a man called Charles Byrne, who eked out a living at fairgrounds and freak shows calling himself the Irish giant. Byrne was seven foot seven inches tall, if not the eight feet claimed on the posters. And John Hunter badly wanted him for his collection. The story goes that Byrne was so afraid of being dissected after his death that he asked his friends to bury him at sea in a lead coffin, but his friends betrayed him for money. Whether this is true or not, we don't know, but certainly we know that Hunter paid £130 for the body – in today's terms about £15,000. And we know that he took the corpse, dissembled it, boiled it up in his great cauldron and then reassembled the bones for display.

Now at the Hunterian Museum we can see him: Charles Byrne, the Irish giant who died at the age of 22. He had a pituitary tumour – *acromegaly* – which is why he grew so large with massive amounts of growth hormone. But for me, John Hunter was a true hero because he influenced so much surgery that we do today. So many of the things that he dissected directly influenced how we handle blood vessels and how we handle all sorts of organs when we conduct regular operations.

Robert Winston

One of Hunter's greatest surgical triumphs came in 1785. By then, he was Europe's leading surgeon and anatomist, a fixture in London society, surgeon extraordinary to George III and married to a beautiful poet, Anne Home, who would entertain London's literati in their salon while Hunter, uninterested in literary debate, received deliveries of corpses from London's sack 'em up men downstairs at their back door.

In October 1785, a ship's rigger called John Burley turned up at St George's Hospital – now the Lanesborough Hotel on Hyde Park Corner – with a 9lb facial tumour twice the size of his head. At 37 years of age, he'd achieved the average life expectancy of a person in Georgian England. At this stage of his life, he must have been truly desperate to risk the terrors of the operating theatre, but no other surgeon would touch him. Word had reached him that Hunter, now chief surgeon at St George's, was different to the others because he had made a science out of anatomy and could therefore go where no other surgeon dared to cut.

Burley had come to the right place and right man. Hunter's meticulous technique enabled him to remove the tumour in twenty-five minutes flat, writing in his diary that 'the man did not cry out once'. Every other surgeon had turned Burley away because they did not know where in the tumour they might encounter the facial nerve. Cutting the nerve would have left Burley's face paralysed. From his anatomical studies, Hunter knew to make his incision sufficiently far back on the face – along a vertical line in front of the ear – to avoid the facial nerve. Burley was left with a long, thin scar but no trace of facial paralysis. His eyelid did not droop and his mouth was straight. It was a remarkable result.

Hunter published a succession of texts on his anatomical discoveries and the surgical techniques he had developed, starting in 1771 with *A Treatise on the Natural History of the Human Teeth*, in which he reported his transplantation of a human tooth into a cock's comb. Other texts covered the senses, tissue grafting and various diseases, but his most famous and controversial treatise was published after he performed a somewhat reckless experiment by inoculating himself with gonorrhoea. Unaware that the needle was also contaminated with syphilis, his discovery of the symptoms of both syphilis and gonorrhoea confused him into thinking that there was only one venereal disease.

Hunter treated himself with mercury and cauterisation – hence the saying 'a night with Venus, a lifetime with mercury' – and publicised his findings in his *Treatise on the Venereal Disease* in 1786. Given Hunter's reputation, his findings were accepted without question. His treatise became a bestseller and the true nature of gonorrhoea and syphilis was not discovered until more than fifty years later.

However, this was one small slip in a distinguished career that came to an end in 1793 during a heated meeting with colleagues at St George's Hospital. Arguing over the admission of two students who, like Hunter fifty years earlier, had arrived in London with the dream of becoming surgeons, Hunter stepped out of the meeting to escape the wrath of a disagreeing colleague. In a neighbouring room, he had a heart attack and died.

Hunter left huge debts, having spent most of his fortune on his unique anatomical collection of more than 14,000 items collected over forty years, including a comparison of monkey and human skulls that he said proved 'our first parents, Adam and Eve, were indisputably black'. Hopes that the nation would buy his collection were rejected by the Prime Minister, William Pitt the Younger, who was at war with France.

'What? Buy preparations?' Pitt exclaimed. 'Why, I have not got money enough to purchase gunpowder.'

Hunter's heirs were left with nothing, but his collection was placed in the care of the College of Surgeons, which had just bought a new property at Lincoln's Inn Fields, three years before it was granted a royal charter at the turn of the nineteenth century. It is still largely intact as the Hunterian Museum at the Royal College of Surgeons.

* * *

Although Hunter brought scientific method to the biological sciences, many other scientists of the era allowed their research to be directed more by pragmatism than a total understanding of the underlying science. At any time, scientific progress can be shaped just as much as any other form of human endeavour by guile or cunning or serendipity or surprising leaps of imagination. And the discovery of vaccination by a young Gloucestershire surgeon called Edward Jenner managed to embody all of these characteristic at once, his hunch saving the lives of billions over the years.

Having studied anatomy and surgery under John Hunter, Jenner had turned down a post as the on-board naturalist on James Cook's second voyage to the South Pacific, preferring instead to set up a practice in his hometown of Berkeley in Gloucestershire. In 1788, as Priestley was nearing the end of his decade in Birmingham, Jenner was celebrating having his first paper published in the *Philosophical Transactions* of the Royal Society. For many months he had been indulging in his hobby of bird watching and, having closely watched a hedge sparrow nest, he'd become the first person to report that young cuckoos routinely eliminated their competitors by pushing out the hedge sparrow young. But Jenner didn't have a chance to bask in the glory of his first publication; a smallpox epidemic was sweeping over Gloucestershire that year and he was busy dealing with the outbreak.

Few people at that time escaped smallpox infection at some point in their lives. Thought to have originated more than 3,000 years ago in India or Egypt, smallpox's worst epidemics would leave one in three dead and the survivors severely disfigured for life. During the eighteenth century, it killed an estimated 400,000 Europeans every year, including five reigning monarchs, and it was responsible for a third of all cases of blindness. More than a century later, in 1855, the historian Thomas Babington Macaulay wrote that

... smallpox was always present, filling the churchyard with corpses, tormenting with constant fear all whom it had not yet stricken, leaving on those whose lives it spared the hideous traces of its power, turning the babe into a changeling at which the mother shuddered, and making the eyes and cheeks of the betrothed maiden objects of horror to the lover.

No cure or antidote existed, but one thing was known about smallpox: a mild dose could be better than none at all. After recovery, the victim would have immunity to reinfection for life. This insight had led the Chinese and Turkish to advocate catching the disease from a mild sufferer in the hope that no more than a light dose would be passed on. Although this policy of variolation was often little better in terms of life expectancy than a game of Russian roulette, the practice was brought to Britain at the beginning of the eighteenth century by the wife of the British ambassador to Turkey, Lady Mary Wortley Montagu, renowned as a writer and the most colourful Englishwoman of her time.

The 1715 smallpox epidemic had killed Lady Mary's brother and scarred her famous beauty. Eager to spare her children similar suffering, Lady Mary had them variolated while they were in Turkey and enthusiastically promoted the procedure on her return to London. She encountered considerable resistance from a conservative medical establishment suspicious as much of her gender as of an 'oriental' procedure, but several doctors, among them Jenner, adopted this relatively crude and risky form of inoculation.

However, Jenner wasn't convinced infection with a mild smallpox strain was the best policy. Since about 1775, he had noticed that there appeared to be some truth in the Gloucestershire old wives' tale that infection with cowpox, which affected the teats of cows, conferred immunity to smallpox. A similar immunity also arose in stable workers who had been in contact with horses that had a disease called 'the grease', an equine equivalent of nappy rash that involved blistering and swelling at the back of the fetlock and pastern of a horse's legs.

Few people ever caught the grease or cowpox, it typically being confined to milkmaids – hence the belief at the time that milkmaids had beautiful skin – or dairy farmers. Jenner noticed, however, that he had never seen anyone previously infected with cowpox develop the slightest sign of smallpox symptoms.

> Faced with the question of whether milkmaids didn't get smallpox because they'd had cowpox, Jenner remembered the maxim of his mentor, John Hunter: don't think, try the experiment. In May 1796, that's exactly what Jenner did. It was only a single experiment, but it was enough to prove that he was right.
>
> Richard Dawkins

Convinced there was a link, he needed to test his theory somehow, and the opportunity came on 14 May 1796, when he came across a milkmaid called Sarah Nelmes suffering from a live infection of cowpox caught from a cow called Blossom. Throwing caution to the wind in the name of scientific enquiry, he took a speck of fluid from a blister on the milkmaid's hand and worked it into two cuts in the arm of James Phipps, a perfectly healthy eight-year-old lad who was the son of his gardener.

Phipps reacted as expected. Within a week he had a slight fever, but nothing worse. This was usual for a cowpox infection.

Two months later, on 1 July when Phipps had recovered, Jenner gave the healthy boy a lethal dose of smallpox – an enormously risky deed considering that Jenner was acting on no more than a hunch. Fortunately for the lad, Jenner was proved correct in his assumption. The boy survived and vaccination was born, the word coming from *vaccinia*, the Latin medical term for cowpox. In spite of having no idea how it worked, Jenner then applied sound scientific principles and insisted on repeating the experiment before reporting it as a success. It took him two years to find another person with an active cowpox infection, but he got exactly the same favourable result.

When presented with Jenner's research, the Royal Society ran a mile, not because of the ethical implications of the experiment, but because the Fellows considered that it was 'so much at variance with established knowledge' that it could damage Jenner's reputation.

> Jenner is always given credit in history for the invention of vaccination, but it had actually been done before. It just wasn't widely known in

Jenner's time. A farmer in Dorset called Benjamin Jesty had also heard the folklore surrounding cowpox and noticed its immunising effect.

Jesty had deliberately infected his whole family with cowpox as a means of protection against smallpox – with success – but he is not normally given credit for inventing vaccination, probably rightly because it was Jenner who brought it to the attention of doctors and really advocated and publicised the idea of vaccination.

The story of Jenner, the milkmaid and the farmer's son has become an iconic story in medical history, similar to that of Fleming and the *penicillium* spore. It conveys the idea in a single, vivid moment in history. Of course, it would have been much more drawn out. A lot of thought and work went into Jenner's work before he publicised it. And he followed his publication with many years of further work, ensuring that it was recorded in a form that would influence and convince the whole medical profession.

Jenner's work led to the whole science of immunology. And Jenner is rightly regarded as the father of the science.

Richard Dawkins

Jenner published privately and within a few years smallpox vaccination had been adopted throughout Europe. Within eighteen months of vaccination being introduced in Britain, 12,000 people had protected themselves against smallpox infection, the royal family among them. By 1805, smallpox deaths in England had been reduced by two-thirds. Nearly 200 years after Jenner's discovery, in December 1979, the World Health Organization declared smallpox eradicated.

In the years following his discovery, Jenner was celebrated widely as a hero. In some German principalities, his birthday was declared a holiday. Even in Britain's most implacable enemy, revolutionary France, Napoleon immediately granted freedom to a group of British prisoners simply because Jenner had signed a petition, saying 'Ah Jenner ... we can refuse him nothing.' Meanwhile in Britain, Parliament overcame its usual parsimony towards outsiders to make payments of £30,000 to Jenner in recognition of his service to the country.

However, Jenner's deification had its limits. The British medical establishment was less than welcoming. When proposed for election to the College of Physicians, Jenner thought his discovery of smallpox vaccination was qualification enough and refused to undergo the usual procedure of examination in the classics of medicine. The College, still wedded to the long-since discredited teachings of Galen and Hippocrates, refused him.

* * *

Through his mentoring of Jenner and his advocacy of evidence-based medicine, John Hunter had played a major role in bringing the principles of the scientific method to the biological sciences and specifically to anatomy and surgery. However, as the nineteenth century dawned, the biological sciences were still several steps away from conceiving their equivalent of a Newtonian universal law that replaced mysticism with rationality.

John Ray's systematic classification had illuminated the ways in which diverse species were interlinked. Joseph Banks's travels had revealed the diversity of life and raised profound questions about evolution. Hunter's discoveries had shown how common features and pathologies were shared between animals and humans. Jenner's work had shown how scientific enquiry could reduce human suffering. And yet the religious orthodoxy of a biblical six days of creation, a divine intelligent design and a young Earth only a few thousand years old was still believed by the majority, including many scientists.

It would take Darwin's *On the Origin of Species* to shatter those illusions in such a devastating manner by providing a universal theory that not only revealed humankind's place in the greater scheme of things, but by also explaining the path of all life on Earth. However, Darwin would not have reached his conclusions without evidence that the Earth was old enough for natural selection to have shaped evolution. That crucial stepping stone to Darwinism was provided by Charles Lyell, a Scottish geologist who became a close friend of Darwin.

Although he is frequently called the father of geology, Lyell in fact contributed nothing fundamentally new to the subject. His great achievement was to explain and to popularise the work of his predecessors, in the process reshaping opinion concerning the history of Earth from one that explained its formation as a series of catastrophic life-destroying events, such as the biblical flood, to one in which it was formed gradually by continuous natural forces, such as volcanic eruptions and subsequent erosion, over millions of years. This had a profound effect on Darwin's development of his theory of evolution by natural selection.

Extremely short-sighted and at times quite odd in his behaviour – Darwin described how Lyell would bend over to rest his head on the seat of a chair while he was standing beside it – Lyell grew up in a comfortably well-off family. Born in Scotland, he was taken south at a young age by his widowed mother to the New Forest, where he became interested in nature and a keen butterfly collector.

Already interested in geology before he went to Oxford, he neglected his study of classics to attend lectures by the eccentric Reverend William Buckland. Besides having a menagerie of wild animals, including snakes, eagles and monkeys that he was fond of sampling at his dinner table, Buckland was a geologist who subscribed to the 'neptunist' view that the Earth's crust had built up in the form of layers, or strata, laid down as silt under water over a long time. By contrast, 'vulcanists' believed Earth was formed through a series of volcanic eruptions releasing successive layers of molten rocked that cooled to form strata. Caught under the spell of geology, Lyell nevertheless went on to study law and was accepted for the bar, but whenever he could find time he pursued his interest in the Earth's crust, travelling extensively to collect data.

During a family holiday in Paris, Lyell met Georges Cuvier, an anatomist who was fiercely anti-evolution, believing fossils to be relics of life forms that existed before a previous catastrophic event, such as the biblical flood. He also met Alexander von Humboldt, a German naturalist who had studied volcanoes and shown how their eruptions led to the formation of igneous rocks.

Lyell did not instinctively agree with Cuvier's or Humboldt's theories, but they set him thinking. When he returned to Britain, he continued his investigations. Eventually he came across the work of James Hutton, a Scottish geologist who more than a century earlier had put forward his theory that all the forces that were currently changing the Earth's surface had been occurring in exactly the same way throughout the planet's history. This also meant that everything that had happened in the past could be explained by events that were still occurring. Lyell was intrigued and investigated further.

Hutton's views were backed up by close studies of rock strata, which showed that the Earth's crust was formed by a combination of compressed sediment and molten rock from volcanoes, subsequently eroded by wind, water and sunlight, exposing successive layers of rock. The implication was that the planet's history was immeasurably long, or as Hutton put it, 'there seemed no sign of a beginning and no prospect of an end'.

Although he was the true father of geology, Hutton faced two problems in receiving recognition for his theories, published in 1785 in *Theory of the Earth*. With the French revolution about to kick off across the Channel, English sentiment was opposed to anything that challenged the status quo (as Priestley would discover six years later in the Birmingham riots), so Hutton's assertion that Earth was millions of times older than the biblical account of its history was not particularly welcome.

However, Hutton's bigger problem was that he was an atrocious writer, described by one of his biographers as 'almost entirely innocent of rhetorical accomplishments'. Just about everyone who attempted to read Hutton's *Theory of the Earth* gave up in the face of its impenetrable prose. Hutton had founded a science but no one knew it.

Hutton's loss was another man's gain, and in this case that gain went to a close friend of his: a mathematician called John Playfair, who later became professor of natural philosophy at the University of Edinburgh.

Playfair *could* write. And having spent many hours listening to his friend expound his theories, he was familiar with Hutton's hitherto unintelligible ideas. In 1802, Playfair published *Illustrations of the Huttonian Theory of the Earth*. An instant bestseller, it was the way that Lyell, in common with everyone else interested in the subject, came to understand Hutton's (literally) groundbreaking work.

In 1831, Lyell was appointed to the chair of geology at King's College, London. That year he published his masterpiece, *The Principles of Geology*, its title a somewhat conceited reference to Newton's *Principia*. No matter that there was little original in it except for Lyell's suggestion that some geological eras could be subdivided into shorter distinct periods; in explaining the work of his predecessors extremely colourfully and lucidly, it did the trick. It established the uniformitarianist account of Earth's formation by extremely slow changes over vast expanses of time, a view that alarmed conservative scientists because it inevitably pointed towards some kind of evolutionary explanation for life on Earth.

More importantly, a first edition was in Charles Darwin's luggage as he sailed in the *Beagle* from Devonport on 27 December 1831.

Chapter 9

DARWIN'S REVOLUTION

On paper, Charles Darwin was an unlikely person to originate the most controversial and far-reaching theory in science. A reticent, perfectionist hypochondriac, the young Darwin showed little academic aptitude and no real interest in the family tradition of medicine and science. After taking an ordinary degree leading to a qualification in divinity at Cambridge, he seemed destined for a quiet life as a country vicar until he was approached out of the blue to join HMS *Beagle* as its on-board naturalist on a five-year journey. It changed his life and the course of history.

Even after he had returned from the circumnavigation, Darwin hardly rushed to publicise the ideas he had amassed while observing and collecting thousands of species. Two decades passed before he felt inclined to publish *On the Origin of Species*, the most significant book in the biological sciences and probably the most important idea ever in science, and he only published it at all because a colleague's work threatened to beat him to it.

Before Charles, the Darwin name was already rich with scientific heritage. His great grandfather, Robert, had met Newton in 1718 after presenting a dinosaur fossil he had found in the Midlands to the Royal Society, of which Newton was president at the time. His grandfather, Erasmus, was a radical freethinker and a very successful doctor who helped found the Lunar Society in Birmingham. Fascinated by steam power, Erasmus Darwin also translated Linnaeus' work on classification of species into English and published books on botany and evolution, arguing that the environment caused evolutionary changes in species that were then inherited by subsequent generations. Of course, Erasmus' grandson would show that it wasn't the environment *per se* that shaped the evolution of species. Instead, the relative abilities of individuals to exploit limited resources within any particular environment caused evolution through natural selection – in other words, those that fitted an environment best had the highest chance of survival.

Darwin's father, another Robert, was a very well-to-do doctor who despaired that his second eldest son (Charles was fifth of six children in all) showed no interest in becoming the third generation of Darwins to go into medicine. 'You care for nothing but shooting, dogs, and rat-catching,' Robert told the teenage Charles. 'And you will be a disgrace to yourself and all your family.' It seemed a fair assessment. Sent to board at Shrewsbury School only a year after his mother died, Darwin was bored by the curriculum and spent his spare time taking long walks, during which he developed a fascination for the natural world, including bird watching. He would collect specimens on these walks and on family holidays in Wales, then research them in the school library or in his father's books.

While visiting his elder brother, Erasmus, who was studying medicine at Cambridge, Darwin read Alexander von Humboldt's *Narrative of Travels*, in which the German naturalist and founder of geophysics described his five-year expedition to Latin America. Like Charles Lyell, the young Darwin found Humboldt's description of collecting vast quantities of botanical, zoological and geographical specimens and data hugely inspiring. His interest in the natural world was growing.

By the time Charles was 15, it was clear to his father that he was showing no interest in learning a traditional curriculum, so his father removed him from school. Hoping that Charles might apply himself to the family business of medicine if he had hands-on experience, he appointed his son as his assistant. Attending to the poor of Shropshire, Charles was allowed to examine patients before his father would check the symptoms and prescribe treatments. After one summer with his son at his side, Darwin's father bundled Charles off to Edinburgh to study medicine and to live with Erasmus, who was now in the final year of his medical training.

Robert Darwin's policy could not have backfired more spectacularly. Erasmus and Charles, who had always been close, made the most of their freedom, Erasmus doing only enough work to complete his medical courses. As for Charles, witnessing two operations (both conducted without anaesthesia, one of them on a child) left him disgusted and mentally scarred for life.

Fearing his father's reaction, Charles remained at Edinburgh, attending natural history classes instead of medicine. Eventually his subterfuge caught up with him. Under parental pressure, he crammed the classics before being packed off to Christ's College, Cambridge, supposedly to study for an ordinary Bachelor of Arts degree (he lacked the mathematics ability for the honours degree) in preparation for a specialist divinity course.

At Cambridge Darwin again rebelled against his father's wishes, attending geology and botany lectures instead of the courses required by his degree. However, this time he was encouraged and supported by dons who recognised his natural ability and his capacity for hard work in a field that genuinely fascinated him. Cambridge was in the midst of a craze for beetle collecting and Charles soon became hooked. Obsessed with winning the student prize (he even employed two locals to catch beetles on his behalf), Darwin once stripped bark from a dead tree and caught a ground beetle in each hand. He was just about to leave with his booty when he spotted the extremely rare Crucifix Ground Beetle. Copying the habits of bird-egg collectors, he popped one of the ground beetles into his mouth to free his hand. The beetle excreted an intensely acrid fluid, which burnt his tongue. Charles was forced to spit it out and lost all three beetles. Undeterred, he continued his quest, eventually catching enough beetles to send records of his finds to James Francis Stephens, a renowned entomologist. Stephens was so impressed he included around thirty of Darwin's discoveries in his *Illustrations of British Entomology*, the first time Darwin's name appeared in print.

Cambridge appears to have been the making of Darwin. With his final exams approaching in subjects he had neglected in order to attend botany lectures, he knuckled down in his fourth year and managed to leave the university in 1831 with an ordinary degree in a subject in which he was not remotely interested, writing afterwards that 'I do not know why the degree should make one so miserable'. Immediately after leaving Cambridge, Darwin hatched a plan to sail to the Canary Islands to study nature at the edge of the tropics. He learnt Spanish, persuaded his father to back him financially, crammed relevant texts on geology and botany, and sought companions. Unable to persuade anyone to accompany him that summer, he postponed the trip until the following year and instead went to study rocks in Wales. When he returned home to Shropshire in August, he found a letter awaiting him. It was from the Rev. John Stevens Henslow, a young botany professor he had befriended at Cambridge. Its contents would change the course of his life.

Henslow's letter passed on a request from Captain Robert FitzRoy of the Admiralty asking for a young, well-educated gentleman with an interest in science to accompany him on a two-year trip to 'Terra del Fuego & home by the East Indies'. The position, which was by invitation only, was 'more as a companion than a mere collector', which had led Henslow to recommend Darwin, he wrote, 'not on the supposition of yr. being a finished Naturalist, but as amply qualified for collecting, observing, & noting any thing worthy to be noted in Natural History'.

Darwin knew this was an opportunity too good not to be seized. It was exactly what he had been seeking. However, first he had to persuade his father to meet the considerable cost of his berth.

His father reacted predictably. A trip to the South Seas was unsuitable for a future minister of the Church, Dr Darwin said.

Charles was about to write to Henslow to turn down the offer when at the last moment his uncle, Josiah Wedgwood (yet another member of the Lunar Society in Birmingham), persuaded Dr Darwin to relent. Once convinced, Dr Darwin stopped at nothing to equip his son and cover all his expenses.

On 27 December 1831, Charles Darwin set sail from Plymouth aboard the *Beagle*, a 90-foot three-masted former ten-gun brig, nicknamed the coffin class because of its low position in the water. Any sense of a triumphant departure was short-lived. Vulnerable to capsizing and swamping by even moderate waves, the *Beagle* duly lived up to its class nickname, returning to port twice because of storms.

On its third attempt the ship departed successfully. Darwin slept in the poop cabin with the mate and midshipman and befriended the crew, who nicknamed him Philos in recognition of his position as the ship's intellectual.

At 26, Captain FitzRoy was not much older than Darwin and expected two things from his on-board naturalist: a reliable scientist and dinner company from someone he could regard as a gentleman in the same social class as himself. His predecessor, Captain Pringle Stokes, had felt so lonely and isolated from his crew that he shot himself through the head during the first voyage of the *Beagle*, so FitzRoy's desire for companionship during a long arduous sea voyage was understandable.

FitzRoy, an ardent biblical creationist who thought Darwin was headed for the ministry, was soon very disappointed to discover that his dining partner held liberal and sceptical views of the religious orthodoxy. The two men frequently argued, particularly over Darwin's strong belief in the abolition of slavery, but otherwise the voyage was a great success.

Darwin was not quite 23 years old and inexperienced as a naturalist, yet he took to the task with gusto. While FitzRoy and his crew performed the geophysical measurements and hydrographic surveying required by the Admiralty, Darwin travelled inland whenever the opportunity arose, riding up to 700 miles on horseback before returning to the ship. In Lima and Montevideo he witnessed revolutions. On the pampas of Argentina he saw gauchos killing natives. In Brazil he experienced the carnival.

However, the most formative experiences were the natural disasters Darwin witnessed and survived at first hand. He saw Mount Osorno on the island of Chiloé in southern Chile erupt. He was walking in some woods in Valdivia, also in southern Chile, when the ground was ripped apart by a large earthquake. These events and other experiences, such as surviving a tidal wave generated by a calving glacier, convinced Darwin of the awesome forces of nature over humankind.

Walking through the ruined town of Concepción, brought down by the Chilean earthquake, he noticed how the tremors had pushed shellfish beds high above the shoreline. Declaring these sights to be 'the most awful yet interesting spectacle I ever beheld', he realised that these were the geological strata revealed by sudden catastrophic events that he had read about as 'incremental upthrusts' in Lyell's *The Principles of Geology*, a gift from FitzRoy that lay beside his bed in his cabin. Here was first-hand evidence of the power of the planet to shape the environment, and a potent indication that its age should be measured in millennia, not in mere thousands of years. Darwin would later draw on this to explain how Earth's great antiquity would over time produce dramatic changes in the environment, which would play a role in shaping the evolution of all living things.

Darwin fastidiously collected specimens, which were shipped back to Henslow whenever possible. They included a rhea – a flightless but smaller relative of the ostrich and emu – which was later used as evidence of the geographical differentiation of species. He also dug up fossils of an extinct gigantic mammal and amassed a huge collection of plants, invertebrates, insects, shellfish and molluscs, birds and small animals. Species that he couldn't package for return to Britain, such as the iguanas and giant tortoises of the Galapagos Islands, were painstakingly documented in his journals.

On 2 October 1836, five years and two days after the *Beagle* set sail from Plymouth, FitzRoy docked her at Falmouth and Darwin stepped ashore a changed man. He didn't know what to make of his many discoveries. He hadn't begun to formulate his thoughts on evolution. But he knew that his future lay in making sense of what he had seen and experienced.

Soon after arriving home, Darwin met Lyell and the two men became firm friends. Then he set about assessing his five-year voyage of discovery. First, he distributed his samples among various naturalists and comparative anatomists for analysis. Their verdicts on the various fish, mammals and birds were published over five years from 1843 in *The Zoology of the Voyage of H.M.S. Beagle*.

Next he prepared various papers and lectures on geology that he presented to the Cambridge Philosophical Society and the Geological Society of London, which immediately gained him fellowship of the latter. Then he turned his journals into a report shared with FitzRoy of their findings on the voyage, later republished as the *Voyage of the Beagle* without FitzRoy's frankly less interesting contribution.

While busy with his publications and lectures, Darwin was also sifting through the expert reports on his samples that had started to arrive at his London lodgings. The reports on the fossils and birds were particularly interesting. The fossils indicated that South America was home to several unique species of mammal.

The report by an ornithologist friend, John Gould, on the birds of the Galapagos Islands was even more intriguing. Darwin thought he had collected a mixture of birds related to blackbirds, finches and wrens. But Gould's analysis indicated that they were in fact all species of ground finch, each of which had adapted differently to the conditions on their particular island. For instance, the finches on an island with a ready supply of nuts had short and sturdy beaks, while the birds from an island with no nuts had long, thin beaks suited to extracting food from narrow rock crevices. Similar island-specific differentiation was also found in three species of Galapagos mockingbirds.

For the next two years, Darwin tussled with the implications of his research. The Galapagos finches and mockingbirds convinced him that species were transmuting into new species and that the characteristics of each new species depended on its particular environment. But Darwin was baffled by a crucial question: did the environment cause animals and plants to change, and did they then pass on those changes to their offspring? If so, did they force those changes on their offspring while they were in the womb? Or was there some kind of organising energy that shaped the features that animals and plants developed as they grew? With no awareness of genetics and very little understanding of the mechanisms by which traits could be inherited, Darwin was contemplating concepts far in advance of contemporary scientific understanding.

Filling notebook after notebook, Darwin pressed on. A visit to London Zoo in March 1838 made a deep impression when he noticed that the expressions, emotions and behaviour of an orang-utan were 'just like a naughty child'. It convinced him that the gulf between man and beast was narrow, a highly controversial proposition at a time when the Church taught that only humanity had a soul.

The only interruption to Darwin's intense working pattern came when he paused to consider whether to marry Emma Wedgwood, a first cousin. Totally consumed by his research, he applied his working methods even to this decision. After conducting a detailed cost-benefit analysis, he decided the material advantages of becoming engaged outweighed the disadvantages. They announced their engagement in the spring of 1838 and they were married by another cousin the following January.

While engaged and contemplating the start of his own family, Darwin came across a text that suddenly brought a new perspective to the questions that had been running around his mind. In September 1838, while reading books on human statistics, Darwin picked up a copy of *An Essay on the Principle of Population*, written forty years earlier and published anonymously by Thomas Malthus.

Darwin's voyage to South America had provided plenty of evidence that the behaviour and characteristics of animals and plants were shaped by their environment. He had the mechanisms, but he was still looking for the missing part of the puzzle, the reason why those behaviours and characteristics would be passed on to subsequent generations.

Malthus warned that, if left unchecked and unrestricted, the human population would double every twenty-five years, ultimately with cataclysmic consequences. In those words, Darwin discovered the necessary motive for evolution. When the population of any species grew too large for the availability of local resources, the members of that species would have to compete, and two things would happen. Those members whose behaviour or characteristics gave them better access to limited resources would survive. Meanwhile those that already had or subsequently developed unfavourable characteristics or behaviour would perish. The individual changes would occur randomly, but the result over time would be the formation of a new species that 'fitted' best to its environment – hence survival of the fittest.

On 28 September 1838, Darwin noted this insight in his notebook. Equipped with a theory, he now needed to investigate the evidence to support or destroy it.

'Being well prepared to appreciate the struggle for existence … it at once struck me that under these circumstances favourable variations would tend to be preserved and unfavourable ones to be destroyed,' he wrote. 'Here, then, I had at last got a theory by which to work.'

Darwin went on to examine how agricultural breeders and gardeners selected favourable traits in livestock, dogs and plants. By the summer of 1839 he felt his work was complete.

Then, busy with starting a family while publishing work on corals and bumble bees, and suffering ill health most likely brought on by overwork (he complained of flatulence, nausea and stomach troubles), Darwin got on with his life.

He extended his theory of evolution by natural selection into a 35-page sketch in 1842 and two years later into a 50,000 word manuscript for publication after his death, but showed no other signs of wanting to publicise his theory, not least because its incendiary implications would be seen as deeply threatening to conventional beliefs that the ancestry of humankind was entirely controlled by a divine creator.

Darwin's reticence to publish was prescient. In 1844, an anonymous author published *Vestiges of the Natural History of Creation*, in which it was suggested that humans evolved from apes. It was roundly decried by the Church, academics and the media. The world, it seemed, was not ready for a theory of evolution, from Darwin or anyone else. And Darwin was too timid to risk the wrath of the majority.

For the next fifteen years, Darwin devoted himself to his large family while compiling exhaustive studies of barnacles and racing pigeons, and dealing with his declining health, the result of either acute hypochondria or some obscure tropical disease. Whatever the cause, no amount of snuff, bracing cold water cures, electrotherapies or other faddish treatments was able to alleviate the symptoms, which at times left Darwin unable to work for longer than ten-minute stretches without lying down. Fortunately he could survive on the income from his late father's considerable estate.

Given Darwin's health and the lack of a financial imperative, it is perhaps unsurprising that he left his work on natural selection untouched except for efforts to organise his many notes into a coherent state, in which they most likely would have remained until after his death, had a letter not arrived in 1858 from an acquaintance travelling in the Far East.

* * *

That acquaintance was Alfred Russel Wallace. Like Darwin, he was a naturalist. And like Darwin, he had developed a theory of evolution by natural selection, which was so similar to Darwin's own ideas that Darwin later wrote: 'I never saw a more striking coincidence. If Wallace had my manuscript sketch written out in 1842, he could not have made a better short abstract.'

Although Wallace reached the same conclusion at exactly the same stage in his life as Darwin (the younger Wallace was 35 in 1858, the same age as Darwin in 1844), Wallace's route to his discovery had been quite different.

One of three brothers born into a modest and large family, Wallace lived in five homes during his childhood, establishing a pattern of insecurity and restless movement that would characterise his life. His father, in a common middle-class Victorian trajectory, speculated unwisely and the family's finances steadily crumbled. As a result, Wallace had a rudimentary education before having to leave school early to make a living as a surveyor.

At the time that Darwin was returning with his specimens from his voyage on the *Beagle* to his comfortable home in Shrewsbury, the 13-year-old Wallace was lodging with his brother, a builder, in London, his schooling terminated, his future uncertain.

Needing to make a living, the Wallace brothers left London and for a time they based themselves back in Wales, near Neath. The life suited Alfred Wallace, a gangly, awkward but physically tough youth. He and his brother roamed the country, staying in inns and on farms, enjoying the open air, their simple lunches of bread and cheese eaten under a hedge. He learned to measure and to use a sextant, and he began to survey nature.

Slowly, Wallace trained himself as a naturalist. After reading Darwin's advice in *Journal of the Voyage of the Beagle* that travellers should be botanists, he studied plants and created his own herbarium. A friend introduced him to the wonderful world of beetles and he began to collect them. He read *A Treatise on the Geography and Classification of Animals* by William Swainson, building up a picture of species, genera and orders, and becoming interested in questions concerning the differences between species and how they had come about.

While Charles Darwin settled down to the painstaking work that would lead to *The Origin of Species*, Wallace continued to be buffeted by fate. He spent some time as a schoolmaster in Leicester, continuing his self-training as a naturalist and reading Lyell on geology and accounts of travels in South America. After various setbacks – including a recession that temporarily put him out of work, and the deaths of his father and brother – Wallace returned to Wales, where the booming railways required surveyors. At last Wallace had got his life on a secure footing, but he was not inclined to stand still. Wallace had devised a plan to abandon his surveying business, which he found frustrating, and mount an expedition to South America.

> Wallace was nursing a big idea. He knew there was a growing market for exotic bird and insect specimens in Victorian Britain. Why shouldn't he become a collector on a bigger, more adventurous scale, and at the same time continue his researches into the origin of species? He saved all he could from his railway work and researched the flora and fauna of South America. He collected equipment and letters of introduction and, in April 1848, aged 25, Wallace and Bates embarked from Liverpool for the Amazon.
>
> Richard Dawkins

In May 1848, Wallace arrived in Brazil with Henry Bates, a naturalist and entomologist he had known for several years. For a few weeks they explored the Amazon together. Then they decided to go their separate ways.

Wallace spent four years in the jungle, conducting a series of perilous collecting expeditions, recruiting local help as he went. He explored the Rio Negro and ranged as far as the border with Venezuela. He collected butterflies, beetles and birds, monkeys, a small alligator, even a boa constrictor whose hissing 'sounded like high-pressure steam escaping from a Great Western locomotive'. And he recorded the species he found, financing his expedition by sending back specimens to European collectors. It was a hard way to make a living and survival was not easy, particularly after Wallace's younger brother Herbert, who had joined him in 1849, died of yellow fever.

Wallace never forgave himself for his brother's death, and after four years in the jungle he had had enough. He had collected a huge range of flora and fauna for sale to collectors and for his own personal cabinet, including some monkeys and birds that he would take back alive. Pleased with what he had achieved, he wanted to go home.

Three weeks out from Brazil, Wallace's ship caught fire. Grabbing a few drawings and a couple of shirts before taking to a lifeboat, he watched as his monkeys and parrots perished in the flames. His journals, most of his drawings and nearly all his irreplaceable specimens went down with the ship. Wallace spent ten days and nights in an open boat with some companions before being picked up by a passing cargo ship that was in such a poor state that it only just managed to carry them home to England without a further mishap.

Although he lost nearly all his specimens in the sinking, Wallace was fêted as a conquering naturalist back in London. Stoical in his penury, he wrote to a friend that 'You will see that I have some need of philosophic resignation to bear my fate with patience and equanimity.'

Wallace recouped some of his losses through an insurance claim and by publishing a book about his travels, *A Narrative of Travels on the Amazon and Rio Negro*, in 1853. With descriptions of the botany, zoology and physical geography, the book included Wallace's impressions of the native peoples whose existence, completely unaffected by European influence, had made a deep impression on him.

Like Darwin, Wallace noticed that physical geography influenced the species in particular areas and that the species in any particular locale were influenced more by how well they fitted into their environment than by how similar they were to other species. He also published several research papers, including a study of Amazonian butterflies that caught the eye of Darwin, safely tucked up at Down House in Shropshire in the middle of a three-year study of barnacles that helped to cement his reputation.

No closer to publishing his ideas on the evolution of species formulated ten years earlier, Darwin wrote to Wallace shortly before he left for the Far East in April 1854 to ask him to send him any interesting specimens he found on his journey.

Wallace's exploration of Malaysia and Indonesia was extensive and very successful. Stopping at every significant island in the area over the course of eight years, he sailed more than 14,000 miles, collected more than 125,000 specimens and discovered more than 200 new bird species and 1,000 new insect species. He pursued orang-utans and captured birds of paradise. Many of his samples were sent back to customers in Britain, among them Darwin.

While travelling, Wallace made copious notes, which he used as the basis of more than forty scientific papers published while he was still in the Far East. One of his most significant observations was the marked difference between Asian and Australasian species, which he attributed to the impassability of a deep-water channel that separated Borneo and Celebes from Bali and Lombuk, a division now known as the Wallace Line.

In 1855, a year after he had departed, Wallace's first thoughts on evolution were published in the *Annals and Magazine of Natural History*. Wallace put forward the idea that every species had evolved from another very similar species and that, over time, this diversification of species spread like the branches of a tree. Wallace suggested that this implied that all existing species could be traced back up their branches to a common single ancestor.

In London, Charles Lyell, whose work on the age of Earth had influenced Wallace as much as it had Darwin, immediately noticed the paper and brought it to the attention of Darwin, who at that time was in the process of trying to pull together his notes and thoughts on evolution into a coherent state. But Darwin snootily dismissed Wallace's paper, saying it contained nothing new or impressive. It was not long before Darwin revised his opinion of Wallace's ideas.

Three years later, when Wallace was in the Moluccas, he thought back to the Thomas Malthus essay that he had read before his South American expedition. Its account of the checks that limit human populations, such as war, famine and disease, must also act on animals, Wallace realised.

Feverish in his sickbed, attempting to recuperate from a bout of malaria, Wallace came to the same conclusion as Darwin had more than a decade earlier: those individuals who best fitted their environment were most likely to survive long enough to reproduce. Over the course of many generations, the characteristics that made them fittest, like strength, speed or cunning, would become established and the unfit would perish. Eventually a new species would emerge.

In two hours Wallace had sketched out his theory. Over the next two evenings he wrote it out in full in an essay, *On the Tendency of*

Varieties to Depart Indefinitely from the Original Type. He sent it directly to Darwin.

'Those that prolong their existence can only be the most perfect in health and vigour,' Wallace wrote in a conclusion that linked a species' fight to survive with its rate of reproduction and its access to suitable food. 'The weakest and least perfectly organised must always succumb.'

Unaware of Darwin's work on evolution by natural selection, Wallace innocently asked his older friend for his comments.

Wallace's essay hit Darwin like a bombshell. Lyell had warned him that he might one day be pre-empted if he did not publish his theory and findings. Now it looked like that day had arrived.

'Your words have come true with a vengeance – that I should be forestalled,' Darwin wrote to Lyell before commenting on the 'striking coincidence' between Wallace's essay and his own work. Then, ever the gentleman, Darwin suggested that he step aside. 'I shall, of course, at once write and offer to send it to any journal.'

Darwin's apparent willingness to cede priority to Wallace may have been partly motivated by the personal tragedy of his youngest son falling critically ill to scarlet fever; he died within ten days of Wallace's essay arriving on Darwin's doormat. However, Lyell was unwilling to allow another man's work to strip Darwin of the recognition he felt Darwin deserved and promptly came to his rescue. With Joseph Hooker, a mutual botanist friend, Lyell approached the Linnean Society to suggest that Wallace's and Darwin's papers should be presented jointly to the public at an upcoming meeting. Neither Wallace nor Darwin attended. Wallace was still in the Far East and Darwin was burying his son.

On 1 July 1858, the grand theory of evolution by natural selection was presented to the world. It was the greatest moment in science since Newton's publication of *Principia Mathematica*. It was also the most profound and far-reaching explanation of humankind's origins and our position in the universe. But hardly anyone noticed or took interest.

Squeezed between six other presentations on topics such as the flora of Angola, the first presentation of the theory of evolution, which effectively abolished God's influence and changed the world forever, passed without so much as a murmur. A friend of

Darwin's reported 'the interest was intense, but the subject was too novel and too ominous for the old school to enter the lists without armouring'. The only response came later from a professor in Dublin, who read Wallace, and Darwin's papers and commented that everything contained in them was either already known or plain wrong.

Even by the year's end, the implications of evolutionary theory had not really sunk in. Looking back over the year's events the President of the Linnean Society noted that 'there wasn't really anything of significance to report'.

In Borneo, Wallace was now embarked on another series of expeditions, his main quarry the fabulous and very valuable Bird of Paradise. When he heard about the events at the Linnean Society meeting, he was thrilled to hear that his paper had been presented publicly, even though Lyell and Hooker had not sought his permission and had presented his essay after Darwin's paper.

'I not only approved,' he wrote to his mother, 'but felt that they had given me more honour and credit than I deserved.'

Wallace was as generous, modest and self-deprecating as Darwin. He was delighted to be associated with his older acquaintance and did not resent Darwin's work being presented with his own. Far from it, in fact. From then on, Wallace referred to their joint theory as Darwinism. He even went on to champion Darwin against his opponents and to write a book entitled *Darwinism: An Exposition of the Theory of Natural Selection, with Some of Its Applications*.

For me, Wallace deserves to be honoured not only because he had the intellect and imagination to come up with the theory of evolution by natural selection independently of Charles Darwin, but because his behaviour over authorship of that great idea showed such grace and gentlemanliness. To me it exemplifies the best of British science. He was a brave and original man. His life was adventurous and difficult, and he carved it out himself without the benefit of any social or financial advantage.

When Darwin received Wallace's letter his reactions were of amazement and despair. It was as if he were reading his own theory; a theory that he, of course, hadn't published.

Darwin was paralysed. What could he do?

Honour dictated that he make Wallace's paper public. To make matters worse Darwin's baby son was seriously ill with scarlet fever and not thought to live.

Fortunately Lyell came up with his solution of presenting them both at the Linnean Society, but Wallace never evinced any jealousy or resentment at not being hailed as the originator of the theory of evolution. He said that his insight, however important, couldn't really compare with the depth and authority of Darwin's theory, based as it was on years of observation, study and analysis. He entered into an extended and warm correspondence with Darwin and dedicated future books to him. He even coined the word 'Darwinism' and called himself 'more Darwinian than Darwin'.

For me this whole episode is heartwarming. The history of science is littered with bitter disputes about priority. We only have to remember the feud between Newton and Hooke to be reminded how badly scientists, however brilliant, can behave. In the case of Wallace and Darwin both men acted with decency and grace, each prepared to sacrifice personal glory for the advancement of scientific truth.

In his autobiography, *My Life*, Wallace is philosophical: 'I think therefore that I may have the satisfaction of knowing that by writing my article and sending it to Darwin, I was the unconscious means of leading him to concentrate himself on the task of drawing up what he termed an "abstract" of the great work he had in preparation, but which was really a large and carefully written volume – the celebrated *Origin of Species*, published in November 1859.'

So, perhaps, no Alfred Russel Wallace: no *Origin of Species*. And without Wallace in the mid-1800s the world wouldn't have shifted on its axis in the way it did.

Richard Dawkins

Wallace returned to England in 1862. For the next fifty years he wrote countless papers, articles, books and letters on subjects as various as his experiences in the Far East, the behaviour of swallowtail butterflies, female suffrage and socialism. More outgoing than Darwin, he travelled widely to lecture on Darwinism, so that by the time of his death in 1913 he was one of the best-known scientists in the world, and widely regarded as the grand old man of science. In his later years, he diverged slightly from a purely Darwinist interpretation of evolution, believing that while man's physical form was entirely the result of evolution by natural selection, his mental capabilities were the result of a divine 'metabiological' force. These beliefs and an attachment to spiritualism and extra-terrestrial life damaged his scientific reputation so that by the mid-twentieth century his valuable contribution to evolution by natural selection had been largely forgotten.

* * *

Wallace was quite justified in writing towards the end of his career that he considered his greatest achievement to be his prompting of Darwin's publication of *On the Origin of Species*. In November 1859, some seventeen months after his theory was first publicised at the Linnean Society, Darwin's *On the Origin of Species by Means of Natural Selection, or the Preservation of Favoured Races in the Struggle for Life* appeared in bookshops. The title might have been unwieldy, but the contents were stunning in their lucidity.

By cutting down his 250,000 word manuscript to a more manageable 155,000 words, Darwin had turned an academic treatise into a very readable mass-market book that sold out its initial print run of 1,250 copies on its first day of publication. Had Darwin had his own way, *Origin of Species* would have been five times as long and he often referred to the published book as a mere abstract of his intended work.

The book was immediately reissued in a succession of editions that gave Darwin several chances to revise his text, but his last-minute dash to the printer's deadline had left him feeling 'as weak as a child' and he retired with his usual set of ailments – stomach pains, vomiting, rashes and 'fiery boils' – to Ilkley Wells House, a spa on the Yorkshire Moors, to await the public reaction.

Although lauded by many of his contemporaries, including Lyell and Hooker, Darwin was heavily criticised by many other intellectuals. Anti-Malthusians such as Karl Marx and Friedrich Engels objected to the economic and social implications of a theory that saw society in terms of a competition for resources in which only the fittest would prosper.

Several leading scientists questioned the absence of fossil evidence to support Darwin's statements (some of those missing fossils were subsequently discovered). Some scientists refused to believe that the Earth had existed long enough for Darwin's suggested evolutionary changes to take place, while others could not understand how complex organs such as the eye could evolve in slow, gradual steps over time. And many people were simply extremely shocked that Darwin's theory contradicted divine design and the teachings of the Bible. As for the Church, liberal Anglicans strongly supported Darwin's natural selection as an instrument of God's design, but the Church of England as a whole, like the traditional wing of the scientific establishment, reacted equally strongly against the book.

Partly because of his physical frailty, Darwin did not enter the controversy. He later addressed the even more contentious issue of humankind's descent from lesser primates in *The Descent of Man*, but at the time that *Origin of Species* was published he retreated quietly to let his supporters and detractors argue the issues among themselves.

* * *

Fortunately, Darwin had an extremely vociferous cheerleader in the form of Thomas Henry Huxley, a biologist and close friend who took it upon himself to become Darwin's bulldog.

Self-taught to the level of being able to enter medical school, Huxley worked first as a ship's surgeon. During a four-year voyage in the South Seas, he spent most spare moments gazing down a microscope at molluscs and marine invertebrates. He soon realised that the existing classifications of the animal kingdom were inadequate to describe many anatomical distinctions that he had noticed, so he rewrote the classification system. On his return to Britain, these discoveries brought him a Fellowship of the Royal Society, and the post of professor of natural history at the Royal College of Mines, the forerunner of the Royal College of Science.

However, Huxley's much more significant contribution to science came not long after he read *Origin of Species*, to which he responded in characteristic fashion: 'Now why didn't I think of that?'

Having written several supportive reviews of Darwin's book in newspapers and magazines, and concerned and slightly frustrated that Darwin seemed unwilling and too ill to defend his theory in a public forum, Huxley agreed to debate the issue with Samuel Wilberforce, the Bishop of Oxford. They came face to face at a meeting in 1860 of the British Association for the Advancement of Science.

Wilberforce, known as Soapy Sam for his obsequious manner and oily way of speaking, had been carefully briefed by Richard Owen, a zoologist, dinosaur hunter and one of Darwin's most vehement critics. A discussion about anatomical differences between species soon descended into a dispute over the social and moral implications of evolution. In a hall packed with an audience of 700 people, Wilberforce taunted Huxley by asking him if the apes that he claimed were his forebears were on his grandfather's or grandmother's side.

Rising to his feet, Huxley struggled to hide his disdain. 'If the question is put to me,' he said, 'would I rather have a miserable ape for a grandfather or a man highly endowed by nature and possessed of great means of influence and yet who employs these faculties and that influence for the mere purpose of introducing ridicule into a grave scientific discussion – I unhesitatingly affirm my preference for the ape.'

The audience gasped; the press lapped it up. Wilberforce's ill-conceived cheap shot had backfired spectacularly and Huxley's comprehensive dismissal had provided him with the ideal illustration of the virtues of Darwinism. The exchange was widely reported, albeit with the erroneous quote of 'I would rather be an ape than a bishop' attributed to Huxley.

From that day on, the ascent of rational Darwinism was assured. Huxley went on to devise a method of measuring skulls and he founded the science of craniology, which he used to investigate the recently discovered Neanderthal skull. He also invented the word 'agnostic', which he used to describe his religious beliefs in the wake of Darwinism. However, although he published more than 150 research papers on anthropology, ethnicity, classification and anatomy, among other subjects, his greatest contribution to science was as a teacher and a populariser of the works of other scientists, particularly in providing the wings for Darwinism to fly.

Chapter 10

SPARKS FLY

Darwinism might have been the most important scientific theory of the nineteenth century, but it was closely rivalled by the equally profound and far-reaching discovery of the link between magnetism and electricity.

Scientists already knew how to produce galvanic electricity by chemical reaction, in other words, by using a battery. They also understood frictional electricity; that is, static electricity produced by rubbing a suitable material, such as glass or amber. But until the nineteenth century, no one had realised the fundamental relationship between magnetism and electricity.

By the turn of the century the Industrial Revolution was in full swing and gathering speed. The invention and refinement of the steam engine by Newcomen, Watt and Murdock had transformed the industrial landscape and was starting to make itself felt on communications, in particular steam railways and ships. But the biggest change was to come through the harnessing of electromagnetic power, and the hero of this electric age was a self-taught scientist called Michael Faraday.

Faraday's experiments and discoveries not only proved the existence of the electro-magnetic relationship, but also enabled him to invent the electric dynamo and the electric motor. With them, electricity was at last freed from the shackles of the chemical battery, and successions of further discoveries and innovations followed in their wake.

Perhaps surprisingly, Britain took longer than other countries to exploit the potential of electricity. Faraday was the first to demonstrate its relationship to magnetism, but when it happened,

Britain was already strongly wedded to steam power and, thanks to William Murdock, gas lighting was in widespread use. As a consequence, electricity took off sooner in continental Europe and America than in the country of its discovery.

The story behind Faraday's discovery of electrical induction goes back to the first true scientist, William Gilbert, who in the late sixteenth century coined the term 'electrics' to describe materials such as amber that, when rubbed with fur, became polarised in a similar way to magnets.

In the 1660s, Otto von Guericke, mayor of Magdeburg in Prussian Saxony and inventor of the air pump that inspired Boyle and Hooke, exploited Gilbert's discovery to devise the first electricity generator. His primitive device rotated a ball of sulphur, which when he held his hand against it, emitted sparks of static electricity.

For nearly a century scientists knew they could use devices like von Guericke's to generate frictional electricity at will, but they had no means of storing it. In 1745, two Dutch professors provided the answer with a glass tube partially filled with water and sealed with a cork pierced by a nail. If the nail was held to a frictional electricity generator such as von Guericke's machine, the static electricity charge would accumulate in the water until the nail was pressed against a grounded object, at which point the entire charge would be released at once – as one of the two professors discovered the hard way. When Pieter van Musschenbroek touched one of his first charged Leyden jars with his finger, the sudden discharge of all the accumulated frictional electricity nearly killed him.

The next year, a French abbot called Jean Antoine Nollet provided an even more spectacular demonstration of the possibilities of electricity when he arranged about 200 monks in a long snaking line at the Grand Convent of the Carthusians in Paris. Nollet asked each of his brothers to hold an eight-metre iron wire in one hand and to use their other hand to grab the end of the wire being held by their neighbour. In this way, Nollet formed a continuous monk-wire-monk-wire line more than a mile long. Without warning the monks, Nollet touched the two end pieces of wire to a charged Leyden jar. The simultaneous contortions and grimaces of 200 monks each experiencing a considerable electric shock provided immediate visual proof that electricity travels long distances at a very high speed.

The next breakthrough came after Luigi Galvani, an Italian anatomist, noticed that the dissected leg of a frog twitched when he touched an exposed nerve with a rod charged with static electricity. Intrigued, and having heard of Benjamin Franklin's kite-in-a-storm experiment, Galvani then constructed an arrangement of brass hooks to attach a row of frog's legs to an iron lattice. He laid out his contraption during a thunderstorm. Sure enough, all the frogs' legs twitched throughout the storm, but to Galvani's surprise they started twitching before the storm and continued twitching after it. Realising they would twitch whenever they were touched with two metals, Galvani proposed that it was the result of something he called animal electricity.

Galvani's discovery intrigued a friend of his, Count Alessandro Volta, professor of physics at the Royal School of Cuomo. Volta had already devised a device he called his Perpetual Electrophorus, which combined a frictional electricity generator with a Leyden jar to create a one-stop method of producing, storing and transporting electricity. Now he wondered if Galvani's animal electricity had less to do with the muscle in the frogs' legs and more to do with the two metals.

Volta's discovery that he could generate electricity from a pair of metals just as easily if he replaced the frog's leg with brine-soaked paper triggered a fierce controversy between the adherents of animal-electricity and metal-electricity. 'I don't need your frog,' Volta exclaimed with triumphant scorn when he invented his voltaic pile, a series of copper and zinc strips in a salt solution. 'Give me two metals and a moist rag, and I will produce your animal electricity. Your frog is nothing but a moist conductor, and in this respect it is inferior to my wet rag.'

Although he had crushed Galvani's concept of animal electricity, Volta acknowledged the debt of his discovery to his late friend by calling the electricity produced by his voltaic cell – the first battery – galvanic electricity. Like the electricity from a friction generator, Volta's battery produced electrical charge in a continuous stream, but Volta's opponents maintained that his galvanic electricity was somehow different to frictional electricity. In 1801, a British medic turned physicist, William Wollaston, proved them wrong when he demonstrated to the Royal Society how he had used both types of electricity to split water into hydrogen and oxygen. The only difference was that Volta's pile produced less tension (potential measured in volts) but more quantity (current measured in amperes) than electricity produced by friction and a Leyden jar.

Wollaston later went on to play an important role in inspiring Faraday, but before then he made several scientific discoveries. Educated at Charterhouse and Cambridge University, Wollaston graduated in 1789 with a degree in medicine and immediately set up a private practice and a laboratory to conduct research. While still at Cambridge he had become fascinated with platinum. Heavier, rarer and more inert than gold, platinum had become the glamour metal of the late eighteenth century, but it was notoriously difficult to shape or mould. Wollaston's great breakthrough was to discover a way of making this intractable metal malleable. Keeping his method secret, Wollaston set up a business to work platinum. It made him a fortune and in 1800 he retired from medicine to move from Cambridgeshire to London, where he devoted his time entirely to science.

Withdrawn and shy, Wollaston was obsessed with his work and made many discoveries, including isolating the metals palladium and rhodium. He discovered cystine, one of the amino acid building blocks of protein, and he invented a device called a goniometer that could measure angles between the faces of crystals in minerals.

However, Wollaston's scientific career was characterised by as many near misses as hits. He was probably the first person to observe ultraviolet light, but he did not follow through on his research, so the discovery is attributed to another scientist. Similarly, he was the first to notice a series of black lines separating the coloured bands of light formed when white light is split into a spectrum, something that Newton had missed when he did his work on optics. But again Wollaston did not investigate any further, leaving the discovery to be attributed to the German physicist Joseph von Fraunhofer. Considering that the discovery of these characteristic 'Fraunhofer lines' eventually led to a raft of crucial breakthroughs, including analysis of the composition of the Sun, distant stars and planets as well as details of atomic structure and even the formulation of quantum theory, Wollaston's hesitation cost him a starring role in the pantheon of great scientists.

For the purpose of our story, Wollaston's near miss of 1821 was the most significant. In 1820 a Danish physicist called Hans Christian Oersted had discovered that an electric current running through a wire produced a magnetic field at right angles to the wire. News of Oersted's discovery spread like wildfire through the European scientific community. When Wollaston heard it, he deduced that two things should happen if an electric current produced a magnetic field. First, the wire ought to spin if another magnet was brought near it. Second, it ought to be possible to induce an electric current in a wire by moving a magnet close to it.

Although some of the technical details of Wollaston's assumptions were incorrect, the principles were essentially sound. However, he failed to prove either of his suppositions when, in collaboration with Humphry Davy, he performed some experiments with wires, magnets and batteries at the Royal Institution. The two men gave up, but continued to discuss their ideas about magnetism and electricity. At one of the discussions, Davy's assistant, a young Michael Faraday, was present and joined in. Within months, he had devised his own experiment and in the course of it established himself as one of the true geniuses of science.

* * *

Faraday's achievements were all the more surprising considering his background. Born in 1791, one of ten children of a poor blacksmith, he had a very rudimentary education. By the time he took up an apprenticeship at the age of 14 with a newsagent and bookbinder called George Riebau, Faraday could read but he had no mathematics skills. However, the apprenticeship was his making – if not in the way entirely intended by Riebau. Exposed to books every day, Faraday could not resist peeking inside the covers. What he found immediately fascinated him and Riebau encouraged him to read widely. The turning point came when a customer brought in a copy of the third edition of the *Encyclopaedia Britannica* for binding repairs. In its pages Faraday found the first account he had read of electricity. It triggered a life-long obsession.

If one man switched on the lights of the world, it was Faraday. He became an acknowledged leader of British science, but it took him a long, hard struggle to get there.

Electricity bewitched people in the early nineteenth century. No one really understood it. Some thought it was entertaining. Some thought it was good for you. Some thought it could confer life itself and performed gruesome experiments on executed criminals to try and prove it.

Faraday would come to electricity without preconceptions but with a burning passion to unlock its power.

James Dyson

Hell-bent on self-improvement, Faraday read as many books on science as he could find, including Lavoisier's great textbook on chemistry and *Improvement of the Mind* by Isaac Watts. Subtitled *A supplement to the art of logick, containing a variety of remarks and rules for the attainment and communication of useful knowledge in religion, in the sciences, and in common life*, it was a self-help book of the day that advocated writing letters to hypothetical recipients in order to clarify ideas. Faraday took its advice to heart. In 1810, five years after starting his apprenticeship, he started attending lectures at the City Philosophical Society. Paying great attention to improving his spelling and grammar, Faraday wrote up the lectures in notebooks that he bound himself. He then moved on to conducting experiments of his own that he presented at the society's lecture evenings.

By 1812, the year in which his apprenticeship ended, Faraday had decided to abandon his career as a bookbinder to become a scientist, in those days a vocation without any career structure that was usually pursued by young gentlemen of independent means. However, shortly before he left the employ of Riebau, a regular customer caught sight of Faraday's bound lecture notes from the City Philosophical Society. Recognising Faraday's talent, the customer offered Faraday the highly sought-after tickets he'd bought for each of Humphry Davy's last four lectures at the Royal Institution. Engaged to a wealthy widow, Davy was now about to retire from his professorship. Like Faraday, he came from a modest background and, having now risen to one of the highest positions in British science entirely on merit, he embodied everything that

Faraday hoped he might become. Throughout Davy's four lectures, Faraday took copious notes. Afterwards he added coloured illustrations, then bound them in leather to produce a 386-page account that Ribeau proudly showed off to visitors to his shop, including the customer who had given Faraday his tickets.

Attempting to get a toehold in the world of science, Faraday decided to use his account of Davy's lectures as a calling card at the great institutions of British science. 'The desire to be engaged in scientific occupation, even though of the lowest kind,' Faraday later wrote, 'induced me, whilst an apprentice, to write, in my ignorance of the world and simplicity of my mind, to Sir Joseph Banks, then president of the Royal Society. Naturally enough, "no answer" was the reply left with the porter.'

Undeterred, Faraday then wrote to anyone he thought might give him any kind of job in science, no matter how lowly. Again he was refused by all, but Davy at least gave him an interview after being flattered by Faraday's impressive notes of his last lectures. At the end of the interview, Davy recommended to Faraday that he should 'attend to the book binding' but added a promise to consider him if any position arose in the future.

A few weeks later Davy nearly killed himself in the laboratory. For Davy, who made his name through discoveries that often involved inhaling dangerous gases, it was not the first time he had endangered his health in pursuit of his career. This was the occasion that he damaged his eyesight in an experiment on nitrogen trichloride, the first high explosive, recently isolated by the French chemist Pierre Louis Dulong, who had lost two fingers and an eye in the process. For Faraday, however, Davy's injury was a stroke of luck. Davy now needed an assistant and he remembered young Faraday.

By then, Faraday had moved on from Riebau. Taking temporary leave from his new position as a bookbinder employed by a Mr De La Roche, Faraday worked at the Royal Institution until Davy recovered, then returned to his day job. Having experienced life as an assistant in one of the world's leading laboratories, he desperately wanted to remain at the Royal Institution, but without a private fortune or an influential supporter, it seemed his ambition would be forever thwarted. However, fortune shone his way again a few months later when, for the second time, Faraday profited from someone else's mishap. William Payne, the Royal Institution's instrument maker with a reputation for drunkenness, had brawled in the lecture theatre. Having fired Payne, Davy was now casting around for a replacement and he remembered Faraday. But before calling Faraday for an interview, Davy listened to advice from a trustee of the Royal Institution. 'Let him wash bottles. If he is any good, he will accept the work; if he refuses, he is not good for anything.'

When it came to the interview, Davy went out of his way to discourage Faraday, warning him that 'science was a harsh mistress, and in a pecuniary point of view but poorly rewarding those who devoted themselves to her service'. Nevertheless, Faraday accepted the position without hesitation, moving into rooms on the top floor of the Royal Institution's quarters on Albemarle Street. With a salary of a guinea a week, less than he had earned as a bookbinder, the position lived up to Davy's warning, but Faraday had no regrets. As well as working as a bottle washer, he embarked on what was in effect a second apprenticeship, only this time he was under the best chemist in the land.

Faraday learnt extremely quickly. After years of reading theory, he could now put his ideas into practice. Before long Faraday was suggesting alterations to experiments, then devising entire experiments of his own. The protégé was snapping at his master's heels.

Six months after joining the Royal Institution, Faraday resigned to accompany Davy and his wife on a grand eighteen-month tour of Europe. Officially, Faraday was travelling as Davy's valet – a circumstance that Davy's wife exploited to the full, making Faraday carry her bags and perform menial tasks – but both the men knew his real position was as Davy's experimental assistant.

The tour took them to the leading centres of European science, where they met some of the greatest scientists of the day. Faraday met Volta, the French chemist Louis Vauquelin, the physician André Ampère, who had attempted to use mathematical deduction to prove electromagnetism, and a host of other leading scientists. These meetings and other experiences on the trip turned Faraday from a scientific ingénue into an expert that Davy resentfully came to realise had started to outshine him.

By the time Faraday and Davy returned in 1815, Faraday was among the best informed and most capable scientists in Britain.

He was immediately re-employed by the Royal Institution as a laboratory assistant on a higher salary and with better accommodation, now working with the new professor of chemistry.

Meanwhile, Davy's star was starting to be outshone by Faraday's. Although Davy was approached by the Society for Preventing Accidents in Coal Mines with the commission that led to his design of the miner's safety lamp, there is considerable evidence that Faraday played a significant role in its invention. Most of Davy's manuscripts submitted to the *Philosophical Transactions* on the subject were written by Faraday, and Davy became particularly resentful when his former protégé highlighted some of the flaws in his design.

At the Royal Institution, Faraday's initial responsibility was to conduct research in chemistry. He was often called upon to provide professional advice on a range of subjects from the composition of gunpowder for the East India Company to court cases concerning patents and pollution. As a chemist he made many discoveries, including methods of liquefying gases under pressure, the isolation of benzene (the building block of organic chemistry) and the use of platinum as a catalyst. However, his greatest achievement, in 1821, came relatively early in his career and before most of his chemical discoveries. Thirty years old at the time, Faraday was in fact considerably older than most other leading scientists when they made their greatest breakthroughs – a reflection, no doubt, of his poor education and late start in science. Nevertheless, his discovery was among the most significant in science and immediately sealed his reputation internationally.

By mid-1821 the archive of published work on electromagnetism had grown so large that the editor of *Annals of Philosophy* asked Faraday to summarise it. Faraday, who had been present when Davy and Wollaston discussed their failed attempts to induce a current in a wire by moving a magnet close to it, accepted the commission and set to work, deciding to repeat all the experiments in the process of reviewing them. These led him to postulate that the magnetic field around a wire carrying electrical current is different to that from a bar magnet: it flows in a circle around the wire. Faraday now needed to test his theory.

In early September 1821, Faraday placed a bowl of mercury on his workbench and positioned a magnet vertically at its centre with one pole sticking upwards out of the mercury. He then dipped a copper wire into the mercury at the side of the bowl and connected it to a battery. He connected a second copper wire to the other side of the battery and led it via a switch to a metal frame from which it dangled vertically into the bowl of mercury. Then he pressed a switch to close the circuit. Electricity flowed from the battery. Immediately, the suspended wire rotated around the magnet continuously until he opened the circuit, when it immediately stopped. Switching the magnet around so that its other pole stuck out of the mercury reversed the direction of rotation. With the magnet's north pole uppermost, it rotated anticlockwise; with the south pole uppermost, it rotated clockwise. The experiment proved Faraday's hypothesis of a circular field to be correct, a phenomenon that he called electromagnetic rotation. In showing that it was possible to produce continuous motion from the interaction of electricity and magnetism, Faraday had invented a machine that turned electrical energy into motive force: a simple electric motor.

However, when Faraday reported his findings in October, all hell broke loose. Davy was convinced that Faraday had stolen his ideas from overhearing details of his own and Wollaston's experiments. He accused Faraday of using their ideas without due acknowledgement, but Faraday was able to prove him wrong. Not only had Davy's experiments not worked, but Wollaston (who did not share Davy's anger) had thought incorrectly that the wire would spin along its own axis if a magnet was held nearby, rather than rotating around the magnet as Faraday had shown.

In spite of Faraday's obvious proof, the damage was done and Davy never recovered from being usurped by his former protégé. When Faraday was proposed for fellowship of the Royal Society, Davy attempted to use his presidency to block it. He failed, but it ended their friendship and terminated a fruitful working relationship.

For the next ten years Faraday returned to chemistry, taking up the directorship of the Royal Institution and becoming one of the greatest popularisers in the history of science by inaugurating a series of Christmas lectures for children – a tradition that continues to this day, with lectures now being broadcast annually on television.

Throughout this time Faraday pondered the ultimate implication of his discovery: if an electric current in a wire gave rise to a magnetic field, did it imply that a magnetic field could induce an electric current in a wire? Having used electricity to generate motion in an electric wire, he became obsessed with finding out if it could work the other way – could a magnet be manipulated in such a way that it could produce an electric current? If that was the case, it would be possible to make electricity and to be freed from bulky and short-lived batteries.

In August 1831, Faraday spent around ten days carefully winding two coils of wire on opposite sides of a soft-iron ring and insulating them. When he passed an electric current into one of the coils, an electric current was induced in the other. It was proof that the magnetic field produced by electricity flowing through the first coil had induced an electric current in the second coil. It was also the invention of the transformer.

However, the current in the second coil was induced only for a brief moment whenever the electricity supply to the primary coil was switched off or on. When the current was flowing steadily in the primary coil, there was no current in the secondary coil. It suggested to Faraday that the magnetic field needed to be changing in order to induce a current. Within two months, he had constructed two devices that took into account his previous observations. The first consisted of a wire wound in a coil around an iron helix. When Faraday moved a bar magnet through the centre of the helix, an electric current was induced in the coil. He had achieved electricromagnetic induction, but it produced electricity only as long as the magnet was moving, so the electricity was generated for just a short time. The second device solved that problem.

Faraday attached two wires through sliding contacts to a copper disc positioned to spin between the poles of a horseshoe magnet. When he rotated the disc, he obtained a continuous direct current – the magnetic field was constantly changing, so a constant current was induced. Faraday had invented the first dynamo, and its impact on the world was profound and rapid.

Within a year, the splendidly named Frenchman Hyppolyte Pixii improved on Faraday's design to produce the first effective commercial generator. Five years later, the first working electrical motors were in production. Less than twenty years after Faraday's discovery, private companies were using steam-powered generators to produce electricity for street lighting and theatres, often selling surplus power to neighbouring businesses. In 1881, exactly fifty years after Faraday's discovery, Godalming in Surrey hooked a generator to a water-wheel at a local mill to become the first town in Britain to have a public electricity supply.

Faraday's discoveries would transform the world, but practical applications were never his first concern. For Faraday, electricity came from God and finding out was its own reward. He never patented anything. Once, when asked 'What use is electricity?', he famously replied: 'What use is a baby?'

But Faraday was quite happy to see his discoveries used for the benefit of his fellow man. In the 1850s Michael Faraday was asked to be an adviser to Trinity House, the body that supervised the country's lighthouses. He was convinced that electricity could be used to power the lamps and he supervised the installation of an electromagnetic generator. In December 1858 South Foreland became the first lighthouse to show an electric light.

His legacy was enormous. Before the century was out electric trains were running in Britain, Germany and America. There was electric lighting in London and Liverpool. All over the world people's lives were about to be transformed by the results of Michael Faraday's simple experiments with magnets, coils and iron filings!

Electricity? What use indeed.

James Dyson

Without a mathematical training, Faraday struggled to explain the reasons for electromagnetic induction, so he resorted to his gift for pictorial explanations. Drawing on the familiar image of iron filings sprinkled on a piece of paper above a magnet, he coined the term 'lines of force' to describe a magnetic field. He then explained that it was the movement of these lines of force that, like 'vibrations upon the surface of disturbed water', pushed current through the wire. His suggestion that magnetism was a waveform analogous to sound 'and most probably to light' was a radical departure from the Newtonian concept that forces such as magnetism acted instantaneously whatever the distance. It was also entirely correct.

Faraday continued his research. In 1833, he combined his knowledge of chemistry and electricity to establish fundamental laws of electrolysis. In the process, he coined many of the terms we use today, such as electrolyte, anode, cathode and electrode. Most of them were suggested by William Whewell, a science writer and philosopher who also coined the terms scientist and physicist at this time, writing that 'as an Artist is a Musician, Painter, or Poet, a Scientist is a Mathematician, Physicist, or Naturalist'.

Faraday also investigated the effect of magnetism on light and discovered that it caused the plane of polarisation of light to rotate. This led him to develop an electromagnetic theory of light that was taken up by James Clerk Maxwell and which in its principles laid out a view of the nature of matter that anticipated Einstein's discoveries. Faraday believed that electricity, magnetism and gravity were in essence the same and that their lines of force held the universe together. In this vast soup of forces, matter formed where the forces clumped together – from stars and planets on a galactic scale to atoms on an infinitesimally small scale.

'The view which I am so bold to put forth,' Faraday wrote in a letter to Richard Phillips, a friend and editor of the *Annals of Philosophy*, 'considers radiation as a high species of vibration in the lines of force which are known to connect particles and also masses of matter together. It endeavours to dismiss the aether, but not the vibration.'

Faraday was already in his fifties by this time. Like Newton, he had suffered a nervous breakdown brought on by overwork and was never quite the same after it. However, he continued to put forward ideas that, although not supported by experimental evidence or mathematical analysis, were breaking new ground long after he had achieved more than enough to establish his reputation as a genius of science.

In many ways Faraday was unique in British science. Entirely self-taught, he was the embodiment of science becoming a meritocratic profession rather than a rich man's hobby. In spite of his meagre education, he made discoveries that elevated him into the company of Newton and Darwin. In fact, his lack of mathematical ability helped to shape his genius. It meant that, instead of relying on deduction, Faraday gained his insights through experimentation and a remarkable ability to picture and explain the

I Brunel.

Britain's greatest engineer scientist, Isambard Kingdom Brunel, in front of the launching chains of the ship that killed him, the SS *Great Eastern*. Seen here in November 1857, shortly before her mismanaged launch, she was the largest ship then built. Brunel's beloved Great Babe was a maritime masterpiece, but the logistical and financial stresses of her construction paralysed him and he died within days of her maiden voyage.

Reticent, health-obsessed and perfectionist, Charles Darwin
dithered in publicising his theory of evolution by natural
selection until Alfred Russel Wallace independently arrived at
the same conclusions and threatened to steal his glory.

If he hadn't been so decent and respectful of Darwin, Wallace would have been credited for the discovery of a theory that would now be known as Wallacism, not Darwinism. One of the world's best known scientists when he died, his role as Darwinism's midwife is frequently unappreciated.

A key turning point in Darwin's formulation of evolution by survival of the fittest came when he discovered that each of these Galapagos finches had a beak shape adapted to exploit the environmental conditions and food available on their particular island.

Tenacious, enthusiastic and resourceful, Ernest Rutherford, seen on the right, discovered nuclear radiation, decoded nuclear structure and was first to split the atom, but just as significantly, he inspired and managed a generation of Nobel Prize-winning nuclear scientists at the Cavendish Laboratory in Cambridge.

Ambitious, clever and precocious, William Thomson was a classic Presbyterian Victorian go-getter. He enrolled at Glasgow University at the age of 10, and in his teens moved on to Cambridge. He realised heat was a form of energy that could be converted into mechanical work. This gave Victorian engineers a framework to help them build better machines.

Seen here teaching a physics class at Glasgow University, he was raised to the peerage as Lord Kelvin for laying the first transatlantic telegraph cable.

J.J. Thomson discovered the electron, previously thought
to be a form of radiation, and thereby proved the existence
of particles smaller than atoms – a 'somewhat startling
matter', he told a stunned lecture audience in 1897. It won
him the 1906 Nobel Prize in Physics.

Complex questions generated
by decades of nuclear research
were finally answered in 1927
by Paul Dirac, who united
quantum theory and relativity in
an equation that was the height
of elegance and beauty to anyone
who understood it.

Grandparents and father of the modern computer: Charles
Babbage (left), Ada Lovelace (centre) and Alan Turing (right).
To overcome the human errors involved in compiling tide and
mathematical tables, Babbage developed the first mechanical
computer in 1822. His friend Ada, Countess of Lovelace, wrote
the first computer programme. In the 1930s, Alan Turing invented
the concept of the modern computer – the Universal Machine –
and in the 1940s applied his ideas to breaking German military
codes at Bletchley Park. He is credited with shortening the
Second World War by at least two years.

When Frank Whittle was developing the jet engine, the Air
Ministry was so uninterested it didn't even bother sending a
photographer to its first test flight. Later it realised the folly
of ignoring an invention that could have hastened the end of
the Second World War and commissioned an official film,
Jet Propulsion by the Crown Film Unit. Whittle is seen here
recreating the testing of the engine on its original testbed for
a scene in the film.

Alexander Fleming's serendipitous discovery of the antibiotic properties of the *penicillium* mould would most likely not have happened if he had been more fastidious in cleaning away his culture plates before he went on holiday. When he returned, he noticed the mould's dramatic destruction of *staphyloccus* bacteria. Howard Florey and Ernst Chain later found a way of mass-producing penicillin.

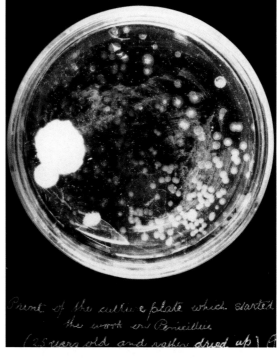

Print of the culture plate which started
the work on Penicillin.
(3 years old and rather dried up)

In 1953 James Watson (left) and Francis Crick, seen here
with a model of part of a DNA molecule, discovered the
double helix structure of the molecule that makes us what
we are. However, their breakthrough relied heavily on
access to the X-ray crystallography of Maurice Wilkins
and Rosalind Franklin at King's College in London. Crick,
Watson and Wilkins shared the 1962 Nobel Prize in
Physiology or Medicine, Franklin having died of cancer.

physical world in clear, lucid terms. As a consequence, his lectures at the Royal Institution eventually surpassed those of Davy in popularity, attracting Charles Dickens and Prince Albert and rescuing the Royal Institution from another of its close brushes with bankruptcy.

Extremely religious, Faraday was unimpressed by honours and uninterested in plaudits. The only recognition he sought was that of his peers through fellowship of the Royal Society. Offered a pension by the prime minister, Lord Melbourne, he felt slighted until he received an apology. He also turned down a knighthood and the presidency of the Royal Society, as well as the opportunity to make a fortune by using his skills to produce poison gas for the Crimean War, responding to the government that although he was capable of manufacturing their deadly intended product he wanted nothing to with it. Modest to the end, his wish for 'a gravestone of the most ordinary kind' and only friends and family at his funeral was honoured when he died at Hampton Court in 1867. His legacy, however, lived on in the Farad unit of capacitance, in the Faraday constant of electrolysis, and in every electrical innovation and machine we use today.

On the infinitesimally small scale, Faraday's discoveries gave impetus to nuclear physicists such as John Joseph Thompson, who discovered the electron, which in turn led to discoveries concerning atomic structure and, eventually, the splitting of the atom. On a larger scale, Faraday's discoveries meant that the electric telegraph, fax machine, telephone and wireless radio ultimately replaced the horseback messenger, semaphore and drum beats as the means of conveying information over long distances.

The repercussions were profound and far-reaching. In a parallel to the first dotcom boom that accompanied the widespread adoption of the internet, the electric age prompted many social changes, for instance the suggestion that invented languages such as Esperanto should be adopted as a means for all people to communicate in a common tongue using the new electrical communications technologies.

Chapter 11

END OF AN EPOCH

Faraday's discoveries and theories fed into a wider debate in nine-teenth-century science about the nature of energy and its rela-tionship with mass. His laws of electrochemistry linked electricity to the forces within atoms and molecules, and his musings on the nature of electricity, magnetism and gravitation suggested they were somehow different varieties of the same fundamental force. Gradually, a generation of British scientists was discovering the interrelationships of electricity, heat, work, energy and mass that would lead to the laws of thermodynamics and conservation of energy.

The first scientist to pick up Faraday's baton was a young brewer from Manchester. Like Faraday, James Prescott Joule had no formal education in science and few mathematical skills, but coming from a wealthy family he had independent means to finance his interest in thermodynamics and electromagnetism. As a shy and delicate child, he started investigating how to construct a more efficient electric motor. By the age of 12 he was publish-ing papers that described the amount of heat produced by electric currents, a relationship now known as Joule's law.

Joule then turned his attention to the relationship between energy, heat and work. He became so obsessed that he took a ther-mometer on his honeymoon. While his bride looked on, Joule measured the heat of the water at the top and bottom of a scenic waterfall they had trekked to visit. To his great satisfaction, he discovered that the water was fractionally warmer at the bottom of the waterfall than the top, thereby proving his hypothesis that the energy of the falling water was converted into heat.

When he returned to his laboratory, Joule continued his investigation of the relationship between heat and work, eventually determining that a particular quantity of heat always produced the same amount of work, and vice versa. This relationship, called the mechanical equivalent of heat, was a crucial stepping stone between the principle of conservation of mechanical energy implied by Newton's laws of motion and the law of conservation of energy, one of the fundamental generalisations in science.

Newtonian physics perfectly described the way in which, for instance, a stone thrown into the air gradually loses kinetic energy of motion as the pull of gravity slows its upward trajectory, but simultaneously gains potential energy as it gets higher. It also explained how the stone's maximum potential energy at the top of the trajectory then decreases as the stone falls back towards the ground and the potential energy of position is converted back into the kinetic energy of motion. However, it took no account of mechanical energy losses due to air friction.

Joule's determination of the mechanical equivalent of heat allowed the increase in temperature of the falling stone and the surrounding air to be regarded as another form of energy. If the stone's changes in temperature and motion, and the air's change in temperature, are all taken into account, then the total amount of energy remains constant.

Joule didn't limit himself to mechanical and heat energy. He also investigated chemical and electrical energy. Then in 1843, at the age of 25, he submitted his findings to the Royal Society and various journals.

Astoundingly, Joule's research was rejected, possibly because he was a brewer rather than a trained scientist with the requisite mathematical skills.

Joule did not give up. He organised a public lecture and coerced the *Manchester Courier*, a newspaper that employed his brother as music critic, to publish his speech in full. The lecture laid out the fundamental findings of his work: that all types of energy produce equivalent amounts of work irrespective of how they are converted into work and the materials involved. The principle was later expressed in the first law of thermodynamics, that energy can be transformed but it cannot be destroyed or created. However, these were complex concepts for a general newspaper readership and again Joule's work went unnoticed.

A few months later, Joule presented his work again in Oxford at the annual meeting of the British Association for the Advancement of Science. Had it not been for a young man sitting in the audience, Joule's work might have been ignored and slipped into the annals of forgotten and overlooked discoveries. That young man was William Thomson, a 22-year-old *wunderkind* who had started attending science lectures at Glasgow University when he was 10 (his father was professor of mathematics). Joule and Thomson became friends and scientific collaborators, Thomson relying on Joule's laboratory expertise while Joule profited from Thomson's mathematical abilities and his standing in the scientific establishment. Together they discovered the phenomenon of cooling in a freely expanding gas, the principle behind the modern refrigerator, now called the Joule-Thomson effect. Independently, Thomson would go on to become one of Britain's greatest scientific geniuses.

* * *

In mid-nineteenth century, Glasgow was the second city of the empire. Its population had trebled between 1800 and 1840. Aside from being a busy port, it was a centre for cotton, chemicals and, in particular, heavy iron engineering. Soon it would send a stream of Clyde-built ships around the world. This was William Thomson's turf, and he was the man who would unlock the huge potential of heat and coal and discover the true nature of heat. His obsession with heat was so all-encompassing that it even determined his leisure activities; all his life he remained a passionate curler.

With a sound mathematical grounding, Thomson had the missing tool needed to turn some of Joule's ideas into laws, and also to develop several of his own theories. However, this imposing and athletic man (although born with a heart problem, he grew into a strong youth and became a keen sportsman at Cambridge University) was also prone to making outlandish and provocative statements that frequently came across as staggeringly ignorant displays of arrogance.

Highly self-regarding of his mental acuities, as a final-year Cambridge undergraduate in 1847 Thomson was so certain that he had achieved the highest mark in the formidable mathematical Tripos exam, a feat that would have earned him the honour of being called the Senior Wrangler, that he issued a terse instruction to his servant.

'Run down to the Senate House, will you,' he said, 'and see who is Second Wrangler.'

Unfortunately for Thomson, the servant returned with the news that the Second Wrangler was 'You, sir.'

Throughout his life, Thomson was no stranger to hubris. In 1895, towards the end of his career, he declared that 'heavier-than-air flying machines are impossible'. Eight years later, the Wright brothers became the first to achieve sustained and controlled powered flight, but even before Thomson uttered his fateful statement men were flying in heavier-than-air gliders, and steam-powered model aeroplanes were flying distances of more than half a mile.

Then, in 1900, Thomson was invited to address the annual meeting of the British Association for the Advancement of Science. To a packed audience of leading physicists he exclaimed that 'there is nothing new to be discovered in physics now. All that remains is more and more precise measurement.'

Five years later, Albert Einstein published his four *annus mirabilis* papers on subjects that fundamentally changed science, including special relativity, matter and energy, space and time, and the photoelectric effect.

However, Thomson's tendency to engage foot with mouth occasionally was the unfortunate flipside of a fervent self-belief that also empowered him to make major contributions to British science. His early work included determining that electrical and magnetic fields spread through a solid in a similar way to that in which heat moved through a conducting solid. He also developed a theory of magnetism.

However, Thomson's greatest contribution to science was to build on Joule's determination of the mechanical equivalent of heat.

The Ulster-born Thomson was a classic Presbyterian Victorian go-getter: ambitious, clever and precocious. He enrolled at Glasgow University aged 10. When he moved on in his teens to Cambridge

University he followed a ferocious work ethic: rise at 5 a.m., read till 8.15, attend lectures, read again, exercise till four, attend chapel till seven, read till 8.30, in bed by nine. If he stuck to it he must have been a machine.

Thomson's big idea was about heat. He realised that it was not a substance, as some thought at the time, but a form of energy, one that could be converted into mechanical work.

He set out how this worked in his Laws of Thermodynamics and these gave Victorian engineers a framework to help them understand, design and build better machines, machines that wouldn't waste energy.

Kathy Sykes

In the mid-nineteenth century, many scientists thought heat was a substance that flowed from areas of high temperature to those of low temperature. In common with Joule, Davy and several other leading scientists, Thomson regarded heat as due to motion or vibration in a material's atomic or molecular constituents. He then provided an explanation of the behaviour of heat that worked with either of the two views. This was his version of the Second Law of Thermodynamics, namely that heat cannot spontaneously flow from cold to hot without work being done.

Although it sounded obvious, the Second Law of Thermo-dynamics had profound implications. It meant that, left untouched, heat dissipated from hot materials to cold materials until all had the same temperature and therefore the same disorder. In other words, the universe over time would 'wear out' from a state of relative order and concentrations of heat and cold to one of uniform temperature and maximum mean disorder. In turn, this implied that everything, the universe included, had a beginning and an end, a concept that contradicted religious eternity and for which there had previously been no scientific proof.

Thomson, who appeared to have some kind of obsession with determining absolutes, used his findings to work out the age of the Sun and Earth and whether there was a minimum possible temperature. He was thwarted in the former, but very successful in the case of the latter. In the case of the age of the Sun and Earth, his mistake was to base his calculation of the age of the solar system on measurements of the rate of cooling of the Earth, which he assumed originated from the Sun.

Unaware of the effects of radioactivity, Thomson did not know that the Earth could maintain its temperature and even warm up, so he vastly underestimated its age. This brought him into conflict with Lyell's theory of gradual geological change and Darwin's theory of evolution though natural selection, so with characteristic arrogance he declared that Lyell and Darwin must be wrong.

As for calculating the minimum possible temperature, Thomson drew on the fact that, for every degree drop in temperature, gases lose one two-hundred-and-seventy-third of their volume at zero Celsius. Thomson made the correct assumption that this meant that the gas's molecules lost one two-hundred-and-seventy-third of their energy for every degree drop in temperature below freezing point, an assumption that in fact holds true for all matter. This meant they would have zero energy at -273 Celsius and therefore it would be impossible to go below this temperature (Thomson's finding has since been experimentally verified, although the exact figure is -273.15C).

Calling this minimum temperature infinite cold – it's now known as absolute zero – Thomson then suggested a temperature scale starting at infinite cold and increasing in increments that matched the degree units of the Celsius scale, so that the melting point of ice was 273 and the boiling point of water was 373. It was later named the Kelvin scale, so the temperatures became 273K and 373K, the name coming from the title that Thomson adopted after his ennoblement as 1st Baron Kelvin.

Thomson was raised to the peerage not for his contributions to science but for work on telegraphy, the first real application of Faraday's discovery of the electromagnetic relationship, and in particular for his involvement in laying the first transatlantic telegraph cable, a project which seized the public's fascination, turning Thomson into a household name when the transatlantic cable shrank the world more than anything before or possibly since.

The electric telegraph was one of the wonders of the nineteenth century, spreading as fast and furious as the internet in the late twentieth century. Only twenty-five years earlier, in 1831, Faraday had invented the electric generator and electric motor, but within three years two Germans, the mathematician Carl Friedrich Gauss and Wilhelm Weber, professor of physics at the University of

Göttingen, had adopted magnetic induction to send messages to a receiver. Four years later, Sir William Fothergill Cooke and Charles Wheatstone installed an experimental telegraph between Euston station and Camden Town in London. In 1843, their telegraph entered commercial use along Brunel's Great Western Railway between Paddington and Slough and was promoted as a tourist attraction at which spectators could witness how 'The Electric Fluid travels at 280,000 miles per second'.

Intended for use as a security and alarm system, the telegraph soon proved its value when, on 1 January 1845, an urgent message was sent from Slough to Paddington:

A MURDER HAS GUST BEEN COMMITTED AT SALT HILL AND THE SUSPECTED MURDERER WAS SEEN TO TAKE A FIRST CLASS TICKET TO LONDON BY THE TRAIN WHICH LEFT SLOUGH AT 742 PM HE IS IN THE GARB OF A KWAKER WITH A GREAT COAT ON WHICH REACHES NEARLY DOWN TO HIS FEET HE IS IN THE LAST COMPART-MENT OF THE SECOND CLASS COMPARTMENT

In spite of being unable to support punctuation, lower case letters or the letters J, Q and Z, the telegraph played a vital role in the apprehension of John Tawell. As soon as he stepped off the train at Paddington, Tawell was shadowed by a detective on to a horse-drawn omnibus and subsequently arrested in a coffee house. Convicted of murdering his mistress, Sarah Hart, Tawell confessed to giving his lover prussic acid after fearing that their relationship was about to become public. He was hanged in public at Ayles-bury.

The case proved the advantages of the telegraph as a rapid means of conveying intelligence and detecting criminals and it was soon adopted on a more extensive scale. It is also believed to be the first use of the telegraph to catch a murderer and the first homicide in which the perpetrator attempted to flee the scene of the crime by railway train.

Meanwhile, Professor Charles Wheatstone, the physicist who had been involved in the Euston to Camden Town and Padding-ton to Slough telegraphs, became a key figure in the spread of the technology, overseeing the laying of a cable to France in 1840. By the mid-1850s he was working for Cyrus West Field, an American financier, on a transatlantic telegraph cable and Thomson had been appointed to the board of directors of Field's Atlantic Telegraph Company.

The company's biggest problem was determining the depth of the Atlantic Ocean, which was so great in some places that the weight of cable required to measure the depth broke the cable reel. Thomson's solution was to develop a small, relatively light device that could be lowered on piano wire. Because of its lightness, the piano wire was influenced by ocean currents, so the depth couldn't be determined simply by measuring the length of wire unreeled into the water. Instead, Thomson's device measured the pressure difference between the surface and ocean floor, from which the depth could then be calculated.

The next problem involved developing a method of picking up the very weak telegraph signal at the receiving end of 3,000 miles of cable. Nineteenth-century technology did not allow for the in-line repeater amplifiers that are now used in long-distance cables, so Wheatstone proposed using large signal voltages to overcome the electrical resistance of the cable's vast length, a suggestion about which Thomson had many misgivings but initially no influence.

In 1857, Thomson, a keen yachtsman, accompanied the first expedition to lay a cable, which was subsidised by the British and American governments. While a converted warship, HMS *Agamemnon,* moved away from a white strand near Ballycarbery Castle in County Kerry on the south-west coast of Ireland, the cable was slowly unfurled. Its seven copper wires were covered with three coats of gutta-percha latex rubber and wound with tarred hemp and a sheath of eighteen strands of iron wire. The cable weighed more than one ton per nautical mile. Unsurprisingly, it broke on the first day.

The cable was grappled and repaired, but it broke again when the *Agamemnon* was 380 miles out to sea over the telegraph plateau, an area nearly two miles deep.

The operation was abandoned until the following summer, when Thomson, as a director of the company, was forced to join the *Agamemnon* unpaid as scientific consultant. He agreed under the proviso that he was allowed to trial a mirror galvanometer he had developed for analysing signals sent over the vast distance of the cable. It could measure tiny fluctuations in electric current. Those fluctuations mattered because those tiny changes in current carried messages in the cables under the Atlantic. Wheatstone, who was also on board, gave Thomson very little opportunity to test his equipment, but Thomson proved his worth in problem-solving under pressure, when the extension to the previous year's cable broke after less than 3 nautical miles, again after 54 nautical miles and for a third time when about 200 nautical miles of cable had been run out.

Eventually, in early August, the cable was completed and the shore end was landed. On 16 August, the first message was sent across the cable. 'Glory to God in the highest; on earth, peace and good will toward men.'

Then Queen Victoria sent a congratulatory telegram to President James Buchanan, expressing hope that the cable would prove 'an additional link between the nations whose friendship is founded on their common interest and reciprocal esteem'.

The president responded that 'it is a triumph more glorious, because far more useful to mankind, than was ever won by conqueror on the field of battle'. The next morning a grand salute of 100 guns resounded in New York City, the streets were decorated with flags, the bells of the churches rung, and at night the city was illuminated. In London, *The Times* declared the link between Valentia in Ireland and Newfoundland no less than 'the most wonderful achievement of this victorious century'. On both sides of the Atlantic on the following Sunday, the transatlantic cable became a theme for innumerable church sermons and a vast amount of hyperbole.

By September, however, the signal was deteriorating. White-house's instruments for receiving the signal were replaced with Thomson's more sensitive mirror galvanometer. Instead of a pointer on a dial, Thomson's device had a moving mirror that in effect amplified the response to the signal by projecting a narrow beam (like an extremely long and weightless pointer) along a scale.

Piqued, and desperate to maintain that his equipment was providing the service, Whitehouse turned the voltage up to 2,000 volts. The insulation failed and it shorted into the ocean. The cable was destroyed.

Reaction to the news was overwhelming, with some writers suggesting the line had always been a hoax, while others declaimed it as a madcap stock exchange speculation.

Following a lengthy parliamentary inquiry, Thomson was put in charge of developing a new link. He redesigned the cable in consultation with Faraday. He then refined his mirror galvanometer, invented a highly sensitive signal generator and a siphon recorder, which operated much like a modern inkjet printer, spraying dots of ink onto paper in response to the signal. With these patented devices, Thomson could send a single Morse character across 3,000 miles every three and a half seconds. He also developed a sensitive measuring system that could monitor the cable at any point.

By 1865 they were ready to go again, but this time they had an important new ally – a great ship, the SS *Great Eastern*, the last ship designed by Isambard Kingdom Brunel.

* * *

Arguably Britain's greatest ever engineer, Brunel was at the very end of his highly illustrious career when he launched the SS *Great Eastern*. The only son of the pioneering tunnel engineer, Marc Isambard Brunel, Isambard started his working life working for his father as resident engineer on the Thames Tunnel – an ambitious project to create a link under the Thames at Rotherhithe. But the job didn't last long. Seriously injured in 1828 by a sudden flood of water into the tunnel, he took several months off to recuperate in Bristol, where he heard about plans to build a bridge to span the Avon gorge at Clifton. After submitting four designs into a competition and battling to overcome the lead judge's preference for his own scheme, Brunel won the commission, although it was not completed until after his death.

Addicted to the impossible, Brunel was not just an engineering pioneer; he made quantum leaps into the unknown. He was audacious, brave and a compulsive risk taker, who never accepted current thinking and always wanted to find a better way. Only in his early twenties, Brunel slung a cable across the Avon Gorge, the first step in his preparations for building the longest single span bridge yet attempted. Edging across it in a basket, the rope snagged. The fearless Brunel climbed onto the wire to free it, 200 feet above the ground. Thankfully he didn't fall, and this obsessive, driven man went on to create so much: bridges, tunnels, railways, even a prefabricated hospital for Crimea. And he always thought big.

James Dyson

Brunel's winning of the commission alerted the burghers of Bristol to his talent. He was then asked to renovate Bristol docks, in the course of which he became fascinated by the potential for railways and persuaded a consortium of Bristol businessmen to petition for a railway to be built from London to Bristol.

In 1833, Brunel was appointed chief engineer for the railway project. When a nervous investor expressed concern at the length of the proposed line, Brunel replied, 'Why not make it longer and have a steamboat go from Bristol to New York and call it The Great Western?'

The name immediately appealed to Brunel, and with characteristic chutzpah he announced it to the public: the Great Western Railway. Then he set about the mammoth construction of a network of tunnels, bridges, viaducts and stations; in all, Brunel oversaw the building of more than 1,600 miles of railway in western England, Wales and the Midlands. No sooner had he finished the railway part of the project than he was grandly announcing the next stage in his grand vision: that passengers would be able to purchase one ticket at London Paddington to travel all the way to New York, changing from the Great Western Railway to the SS *Great Western* steamship at a transatlantic sea port he proposed to build adjoining his railway at Neyland in South Wales.

Of course Brunel ensured that the SS *Great Western* would be the largest ship the world had ever seen. But there were sound engineering principles behind his taste for superlatives. Until then, marine engineers had always insisted that steamships could not cross the Atlantic without the aid of additional sails simply because they couldn't carry sufficient coal. Simply building larger ships was not the answer, the detractors said, because that would increase the drag, which in turn would require more power and consequently even more coal.

At a meeting of the British Association in Bristol, a Dr Dionysius produced calculations which he said showed it was impossible for a ship to carry enough coal to cross the Atlantic. Brunel told the good doctor to 'wait and see' and relied on simple geometry to dismiss his critics. Any vessel's capacity is proportional to the cube of its longest dimension, a law of geometry that applies equally to tin cans and steam ships. For Brunel, the beneficial consequence was that a longer vessel would be able to carry considerably more coal. Meanwhile the water friction that creates drag increases only by the square of the longest dimension. For Brunel and any other shipbuilder this meant the capacity would increase more with length than the drag would increase, so the answer was to build longer and larger.

At 236 feet in length, with four masts to carry sails for auxiliary power and a displacement of 2,340 tonnes, the SS *Great Western* was the largest ship built until that time. Paddle-powered and built in oak, she attracted large crowds when she arrived in London to have her engines installed. However, on the return voyage to Bristol, a fire broke out in the engine room, spreading through the oak deck beams. Rushing to assist in extinguishing the fire, Brunel fell 20 feet off a ladder on to a lower deck and was seriously injured. He was put ashore at Canvey Island while his ship continued to Bristol.

Repairs to the fire damage delayed the *Great Western*'s departure for New York from Avonmouth until 8 April, giving rival steamship operators a chance to beat Brunel in becoming the first to have a ship complete the Atlantic crossing under steam power.

The British & American Steam Navigation Company of London adapted an Irish package steamer, the *Sirius*, for the journey, removing part of her passenger accommodation to make room for extra coal bunkers. Commanded by a naval officer, the *Sirius* sailed from London, reaching America nineteen days later after a very stormy passage, limping into New York harbour after the crew had resorted to burning her cargo, cabin furniture, one of her masts and some decking planks when they ran out of coal. The *Sirius* had achieved the honour of becoming the first steampowered ship to cross the Atlantic, but the triumph was short-lived.

The next day, the *Great Western* arrived. She taken only fifteen days and five hours, with almost 200 tons of coal to spare.

The *Great Western* was a triumph, but it wasn't enough for Brunel. He had an urge: bigger – better – faster. And so in July 1839 he began work on the SS *Great Britain*.

Although Brunel built big – the longest ship in the world – for him, as for all the Victorian engineers, it was all about the efficient use of energy. The speed of a ship is a function of the square root of the length multiplied by a constant of 1.34. So the longer the ship, the faster it would go.

Before the SS *Great Britain*, all ships had paddlewheels. These were relatively efficient, except that in high seas one wheel would often rise out of the water. The ship would slew, creating horrendous loads on the drive system. Brunel reasoned that a single screw propeller, low in the hull, would stay submerged in even the roughest conditions. It would drive smoothly and would be far more efficient. Brunel took the plunge: he committed to screws.

Brunel was not the first person to consider using a screw to propel a vessel through air or water. Leonardo da Vinci toyed with the idea of using it to propel objects through the air, while James Watt was the first to think about using it for ship propulsion. However, Brunel was the first to investigate its potential using scientific methods of testing.

The reverse of the water-raising screw invented by Archimedes more than 2,000 years earlier, the screw propeller is a perfect example of the way the potential of a great idea can be staring people in the face yet be overlooked for centuries. Whereas Archimedes' screw remained stationary relative to its surroundings, moving the water within it as it rotated, the screw propeller was a rotating screw moving through stationary water. Its propulsive force came from the steady flow of mass pushed rearwards, an exploitation of Newton's dictum that every action has an equal and opposite reaction.

By the time Brunel came to the screw propeller, a slew of patents had been granted to various people. In 1835, Francis Pettit Smith carried out experiments with a small boat driven by a wooden screw. In 1838, he launched the first purpose-built screw-driven ship, the *Archimedes*, which sufficiently impressed Brunel to convince him to build the first screw-driven ship to cross the Atlantic. But first he insisted on testing his design until convinced that it was absolutely perfect.

I love the way Brunel investigated and refined. He set up a tug of war between the *Rattler*, a screw-driven sloop he'd chartered and, the *Alecto*, a paddleship. The *Rattler* won.

Brunel then trialled thirty-seven different propellers, changing them a little each time. The results of each tiny iterative change was meticulously recorded by Brunel in his notebooks. Then it was plotted on graphs. For me, this was the beginning of proper research and development.

Brunel eventually came up with a monster six-bladed propeller made of four tons of iron and measuring fifteen feet from blade to blade. It was only 5 per cent less efficient than a twenty-first-century propeller and it propelled Britain in the mid-1900s into the age of transatlantic ocean travel.

James Dyson

With 3,676 tonnes displacement, the *Great Britain* was not only the largest ship to cross the Atlantic but also the first with an iron hull and a screw propeller. It was a huge success, spending more than thirty years sailing to San Francisco and Australia until it ended up damaged in a cove on the Falkland Islands before being salvaged in 1968 and returned to Bristol for renovation.

But even the *Great Britain* was not enough for Isambard Kingdom Brunel. He was consumed by the idea for an even greater ship. She started as *The Leviathan*, became the *Great Eastern*, but to Brunel she was always the Great Babe.

No one asked Brunel to build her. She was his own gigantic idea, a steam-propelled ship with a screw propeller and paddles so huge that she could carry enough coal to sail to Australia and back without refuelling. At 700 feet and nearly 32,000 tons displacement fully laden, she would be twice the size of the *Great Britain*, with capacity for 4,000 passengers and more than 10,000 tons of coal. And she had a double-skinned hull, a great innovation.

Brunel persuaded the Eastern Steam Navigation Company to bankroll the project, but he committed himself heavily, both financially and emotionally. The ship consumed the last years of his life. Too big to be built in Bristol, she was constructed by a Thames yard on the Isle of Dogs, chronicled in a series of remarkable photographs by Robert Howlett. From the outset, the build was troubled. Builders and contractors went in and out of bankruptcy. Funds dried up. A rumour that a welder and his boy had been entombed in the hull created a sense of an unlucky ship.

Her launch in January 1858 was not the low-key event Brunel had intended. In an attempt to recuperate lost money, his backers had sold 3,000 tickets and a circus-like atmosphere ensued. On the rostrum, preparing for the launch, Brunel was asked which of a list of names he preferred. 'Call her Tom Thumb if you like,' said Brunel, clearly disgruntled.

At 12.30 p.m. the daughter of a major backer christened the ship *Leviathan*, much to everyone's surprise as she was commonly known as the *Great Eastern*, then set the launch in motion. However, at the critical moment the giant ship nudged just four feet down the slipway. The steam winches and manual capstans used to haul the ship towards the water were not up to the job. Brunel made another attempt on 19 January and again on the 28th, this time using hydraulic rams to move the ship, but these too proved inadequate. The ship was not finally launched until 1:42 p.m. on 31 January using more powerful hydraulic rams supplied by a new Birmingham company that swiftly capitalised on its association with the biggest marine engineering project the world had ever seen. Six months later, while she was being fitted out, her name was changed back to *Great Eastern*.

Maybe the name change jinxed the ship. Certainly Brunel was running out of luck. The launch had cost £170,000, a third of Brunel's estimate for the entire vessel. Close to bankruptcy, the Eastern Company sold the ship to a newly formed company and the shares lost 90 per cent of their value. A few days before the *Great Eastern* made her maiden voyage, Brunel was photographed on her deck. He looked unwell, stress and his inability to delegate responsibility doubtlessly playing a big part. A couple of hours later he collapsed on board. Struck down with paralysis, he was too ill to join the *Great Eastern* when she sailed a few days later on 6 September 1859. Unfortunately the ship's bad luck followed her. Three days later, off Dungeness, an explosion ripped out her forward funnel, killing five stokers and injuring several others. The news was conveyed to Brunel in his sickbed anxiously waiting to hear of his Great Babe. He died seven days later.

As for the ship that had killed him, the *Great Eastern* made several trips a year to America, but was plagued with technical problems, mishaps and bad luck, including ripping a hole in her hull sixty times the size of the damage suffered by the *Titanic*.

In 1863 the *Great Eastern* made three voyages to New York, with 2,700 passengers being carried out and 970 back to Britain along with a large tonnage of cargo. One of her paddlewheels was damaged on the last outward trip and she completed it using her screw. On the return journey she ran down and damaged the *Jane*, a small sailing ship. The company lost nearly £20,000 on the voyages due to a price war between the Cunard and Inman shipping lines, and ended up with debts of more than £142,000, which forced them to lay up the *Great Eastern*.

A plan was mooted to offer the ship in a lottery. It came to nothing and she was finally offered for sale in January 1864 at the Liverpool Exchange, the bidding opening at £50,000. No bids were offered and the ship was withdrawn from sale, the auctioneer declaring that it would be reoffered for sale with no reserve in three weeks' time. Three men bought her for £25,000, despite the ship being worth £100,000 in materials alone, then chartered her to the newly formed Telegraph Construction and Maintenance Company for conversion to her new role as a cable layer.

*   *   *

Perhaps unsurprisingly, the *Great Eastern*'s run of bad luck continued when she was chartered to assist Thomson in his third attempt to lay a transatlantic cable. Initially, nothing went entirely to plan. The voyage was again dogged by technical problems and had to be abandoned when the cable was lost after 1,200 miles of it had been laid. However, the next year Thomson's team laid an entire new cable in two weeks. Then they recovered the cable they had started to lay the previous year, spliced a new section of cable to it and finished laying a second fully functioning and reliable transatlantic link.

The transatlantic telegraph marked a turning point in world history, bringing near-instantaneous communication to where it had previously taken days or even weeks. Thomson's contributions earned him his knighthood and set him on a path to riches and invention after invention, including a compass that was the first that could provide a true reading of magnetic North on iron ships. He also invented new types of depth-sounding gauges and tide predictors, and pioneered electric light, his home in Glasgow becoming the first house in the world fully lit by electricity. In 1881 he started working with Joseph Swan, who invented the incandescent light bulb, although it is usually credited to Thomas Edison, who improved the technology and proved more successful at marketing it.

Towards the end of his life Lord Kelvin, as he was now known, became increasingly bewildered by the dawning of the second scientific revolution and acutely aware of the failings of the classical physics that he had been so instrumental in creating.

At a celebration of his fiftieth anniversary as Professor of Natural Philosophy at Glasgow University, he shocked his audience when he said:

> One word characterises the most strenuous of the efforts for the advancement of science that I have made perseveringly during fifty-five years. That word is failure.
>
> I know no more of electric and magnetic forces, or of the relation between aether, electricity and ponderable matter, or of chemical affinity, than I knew and tried to teach to my students of natural philosophy fifty years ago in my first session as professor.

It was a rare expression of humility from Lord Kelvin, who in his last years watched as around him a new generation of scientists was making the discoveries of the second scientific revolution – the electron, X-rays, radioactivity, the photoelectric effect and relativity. When he died in 1907, two years after Einstein's *annus mirabilis*, Lord Kelvin's death symbolised the ending of the classical era of physics.

However, the scientist whose discoveries transformed classical physics into the new era of quantum physics was born after Lord Kelvin had been, yet died twenty-eight years before he did. His name was James Clerk Maxwell, the greatest mathematical physicist since Newton and one of the top-tier geniuses of British science. Maxwell introduced the idea of relativity and would have developed special relativity a decade or more before Einstein had he not died young of cancer.

* * *

END OF AN EPOCH

Albert Einstein famously kept three photographs on the wall of his study. Just as he was to become a giant of science and a household name, so also were two of those pictured there: Isaac Newton and Michael Faraday. And like Einstein's discoveries, Newton's and Faraday's achievements were widely recognised.

The third picture showed a man whom Einstein regarded as greater than the other two – the indisputable giant of science – and yet his name and achievements are relatively unknown to most people, even now. This man devised the second great unifying theory of the universe following Newton's law of gravity. Had he not died young, this man would most likely have beaten Einstein by a decade or more to the theory of special relativity. And although Einstein's theories made much of classical physics obsolete, the work of this man in the third photograph remained as valid as ever. So it is perhaps unsurprising that when Einstein visited Cambridge in the 1920s and someone remarked to him, 'You have done great things, but you stand on Newton's shoulders,' Einstein simultaneously dismissed the comment and acknowledged his true hero and inspiration. 'No,' Einstein replied, 'I stand on the shoulders of Maxwell.'

Born the only son of one of the wealthiest and best-known Scottish families, James Clerk Maxwell was a walking contradiction. In appearance he was the classic Scottish laird: cultivated, somewhat old-fashioned, committed to his local community, dressed in clothes that were more practical than fashionable, an expert horseman and a keen countryman. However, in intellectual terms Maxwell was in a class of his own. A colossus of science and the greatest mathematical and theoretical physicist since Newton, his single greatest achievement was his work on electromagnetism, but he also made important discoveries in many other areas, including producing the first ever colour photograph.

From a young age, Maxwell's genius was apparent, although it did him few favours in early life. As a three-year-old he passed his time investigating how doors, locks and keys worked, and how water trickled from a pond along a stream, across bridges and down a drain. Told that a stone was the colour blue, his reply was succinct and direct: 'Ah, but how d'ye know it's blue?'

Inquisitive and insatiably curious – 'What's the go o' that?' he would frequently ask – Maxwell was educated at his family's country estate, initially by his mother, then as his mother fell ill by a tutor who literally beat his schooling into him. Nevertheless, he kept asking questions of the world, inventing an instrument to study the eyes of dogs, which he particularly loved. He also invented the fish-eye lens after apparently becoming intrigued with the structure of the eye of his breakfast kipper.

At the age of 10, shortly after his mother died, Maxwell was sent to the Edinburgh Academy. With a thick Galloway accent, a stuttering aristocratic manner and dressed in old-fashioned clothes and handmade shoes, Maxwell was regarded as an oddity by his classmates, who promptly nicknamed him Dafty and ridiculed his precocious intellect. But before long Dafty showed his great analytical ability. His curiosity in science stimulated by visits with his father to meetings of the Edinburgh Royal Society, Maxwell discovered an original method to draw a perfect oval at the age of 15. When he described it in his first scientific paper, published in the *Proceedings of the Royal Society of Edinburgh*, many of the readers refused to believe that such an accomplished piece of work was entirely the product of so young an author, who incidentally had also already published his first poetry in the *Edinburgh Courant*.

The next year Maxwell moved on to the University of Edinburgh and met William Nicol, a Scottish physicist who invented the first method of producing plane-polarised light. This triggered Maxwell's life-long interest in the nature of light.

Moving on to Cambridge in 1850, Maxwell was taught by some of the world's leading physicists but soon showed he was their equal, a friend remarking that 'I never met a man like him. I do believe there is not a single subject on which he cannot talk, and talk well too, displaying always the most curious and out-of-the-way information.'

In 1854, at the age of 23, Maxwell graduated as Second Wrangler (like Thomson) in the Tripos mathematics exam and was made a fellow of Trinity College. He continued his undergraduate work on colour vision, showing how by varying quantities of the three primary colours he could make any other colour. Maxwell returned to the subject several times in later years. He confirmed that the eye has three types of receptor, each one sensitive to one of the primary colours, and he showed how the absence of one of them leads to colour blindness. In 1861, he crowned his colour work by producing a photograph of a tartan ribbon, the first ever colour photograph. Its three-colour process laid out the principle followed by almost every colour technology that followed, including photography, film-making, television, printing and computer monitors.

Shortly before Maxwell left Cambridge to take up a professorship at Aberdeen, he produced a summary of Faraday's work that not only summarised the contemporary understanding of electromagnetism but, in pointing out what still needed to be discovered, laid out a manifesto for his own future investigations.

Before electromagnetism, Maxwell turned his attention to Saturn's rings, which had intrigued astronomers ever since they were first observed by Galileo. Opinion varied widely within astronomical circles regarding the composition of the rings, but Maxwell showed that the two most popular explanations were impossible and then proposed an alternative that was eventually verified when space probes were sent to the planet in the late twentieth century.

Working purely from theoretical and mathematical principles, Maxwell proved that solid or liquid rings would collapse under the gravitational and mechanical forces acting upon them as they rotated. The only viable structure, he said, was a system of concentric rings of small solid particles orbiting the planet as independent satellites. This interpretation was confirmed in 1979, when Pioneer 11, the first of three NASA and European Space Agency probes, flew within about 12,000 miles of the cloud tops of the giant gaseous planet and photographed its rings.

Immediately after finishing his work on Saturn, in 1860 Maxwell applied his mathematics to another study of small particles, in this instance the myriad molecules that make up gases. Using statistical analysis, he developed a kinetic theory of gases in which the molecules moved in all directions, bouncing off one another so frequently (more than 8 billion times a second for each molecule, according to Maxwell's calculations) that the variation in their velocities and the distances and directions in which they travel are consequently so small as to produce the impression of a smooth-flowing, continuous fluid. When Maxwell analysed the spread of velocities of the molecules and related it to temperature, he could show that the heat of a gas was a function of the average velocity of all its molecules. This was the final evidence needed to destroy the classical theory that heat was a liquid and to prove that instead it was a form of motion.

Maxwell applied his statistical methods to other properties of gases, such as their density, diffusion and viscosity, and also used them to explain the phenomenon by which gases lose heat as they expand. This is because energy is absorbed in overcoming the attraction between atoms and molecules as they move apart. Some years later, in 1871, Maxwell returned to gases to put forward an idea that was to have considerable philosophical implications for the physical sciences, specifically quantum theory.

According to the Second Law of Thermodynamics (and common sense), in normal circumstances heat can flow only from a warmer body to a colder body. However, Maxwell put forward the idea that if two containers of gas at the same temperature were connected, it was statistically possible (although extremely unlikely) that all the faster-moving molecules would move into one of the containers and all the slower-moving molecules would move into the other. This would make one container hotter. If so, this would make the principle set out in the Second Law of Thermodynamics not an absolute certainty, merely an extremely high probability.

The theory was called Maxwell's demon because, to help illustrate it, Maxwell visualised a tiny demon that guarded a door between the two containers. The demon opened the door only to fast-moving molecules going into one of the containers or to slow-moving molecules going into the other.

As well as bringing the Second Law of Thermodynamics into question, Maxwell's demon put forward the idea that it was no longer possible to speak with total certainty of inevitable or eternal outcomes. Instead, descriptions of the physical world would have to be couched in terms of probabilities. Given enough time, random events of chance could theoretically result in highly unlikely outcomes.

Between devising his kinetic theory of gases and putting forward his demon thought experiment, Maxwell left his position in Aberdeen and moved to London in 1860 to take up the position of professor of natural philosophy and astronomy at King's

College. Here he addressed the questions concerning electromagnetism that he had outlined four years earlier in his paper on Faraday. Over the next two years he published four papers, *On Physical Lines of Force*, that investigated electric and magnetic fields. The papers suggested that electricity and magnetism were indissolubly linked (they could not exist without each other) and that they propagated waves that moved at the speed of light, which Maxwell suggested was an indication that light might also be somehow linked to electricity and magnetism.

> Maxwell knew Faraday, elderly now but still working away at the Royal Institution. He was particularly friendly with William Thomson and would tease him over his problems with the transatlantic cable. 'Under the sea, under the sea,' Maxwell wrote, 'No little signals are coming to me.' But some big ideas were coming to Maxwell, ideas that would change the world and secure his scientific reputation for ever. He used a lot of data that came from Faraday's work, but while Faraday did experiments, Maxwell did maths.
>
> He took information about how electric and magnetic fields behaved, added his new ideas about light and translated them into mathematical symbols. He manipulated these symbols and came up with mathematical results that not only explained how things worked now but predicted new phenomena which would and could be found in the real world.
>
> Jim al-Khalili

In late 1864, Maxwell finished his masterpiece, the paper that would put him on an equal footing with Newton and his *Principia Mathematica*. That paper was *A Dynamical Theory of the Electromagnetic Field*. Early in 1865, in a letter to his cousin Charles Cay, Maxwell commented candidly on the work he had recently submitted: 'I have also a paper afloat containing an electromagnetic theory of light, which, til I am convinced to the contrary, I hold to be great guns.'

The comments could not have been more apposite. In one majestic sweep, Maxwell had unified the three realms of physics – electricity, magnetism and light – in a set of four equations that described every facet of electromagnetic radiation (except for certain quantum phenomena that were unknown at that time) and bound them totally together. A dozen or more other theories that had been proposed as explanations of light were instantly made redundant.

In complex mathematics, Maxwell described how the oscillation of an electric charge would produce an electromagnetic field that propagated in the form of waves. The speed of these waves would be equal to a ratio of a measurement in magnetic units to a measurement in electrical units. That ratio was approximately 186,300 miles per hour – the speed of light.

The coincidence was too great. Surely, Maxwell said, it meant that light was part of the same phenomenon as electricity and magnetism: 'This velocity is so nearly that of light that it seems we have strong reason to conclude that light itself (including radiant heat and other radiations, if any) is an electromagnetic disturbance in the form of waves propagated through the electromagnetic field according to electromagnetic laws.'

Maxwell's equations predicted a whole family of other forms of radiation that also travelled at the same speed as light but which had different frequencies. Already ultraviolet and infrared light had been discovered at the fringes of the visible spectrum. Fifteen years later, when radio waves were discovered by Heinrich Hertz, Maxwell's predictions were proved to be correct. And over the next few decades the discovery of X-rays, gamma rays and microwaves provided further proof.

For less mathematically minded audiences, Maxwell described his discoveries in terms of a long rope with one loose end and the other end tied to a fixed point. Shaking the rope up and down sent a wave rippling down the rope, like the electric field produced by an electric charge. According to Maxwell's equations, a simultaneous magnetic ripple ran down the rope at right angles to the electric ripple. At any point along the path of the wave, both the electric field and the magnetic field were constantly changing, albeit at right angles to each other. The speed of the waves passing down the rope was constant – the speed of light – but the length of each wave determined whether it was light, radio wave, X-ray or microwave.

END OF AN EPOCH

People at the time – including scientists like Thomson and Faraday – didn't really understand Maxwell's ideas, or his language. It was as if Einstein had popped up in the 1860s talking about relativity.

And therein lies another reason why James Clerk Maxwell is so important. He created a new language for science, a new methodology for the theoretical physicists of the coming century. Rather than working with coils and magnets, they could work with mathematical symbols. Through mathematics, they could explore ideas, make discoveries and develop theories that they would then test in the real world.

Maxwell didn't just give us radio waves and the mobile phone: he gave us a new way of investigating the world.

Jim al-Khalili

Only 33 years old when he published his theories and equations, Maxwell left King's College two years later to work at home in Galloway. In 1870, he was tempted back into academia when he was offered the professorship of experimental physics at Cambridge, where most of his lectures were unintelligible to all but the very brightest students. At Cambridge he set up the Cavendish Laboratory, which a generation later became the world's leading centre for the study of radiation and atomic structure. He also spent several years editing Henry Cavendish's papers.

Maxwell's legacy was vast. As Einstein said of him, 'one scientific epoch ended and another began with James Clerk Maxwell.' His ideas brought about a revolution in the way that scientists thought about the entire physical world. He introduced the methods that underlie much of quantum theory and he formulated the concept of the butterfly effect of chaos theory, whereby tiny differences in initial conditions can have eventual dramatic consequences.

However, Maxwell died of cancer in 1879, only 48 years old and too young to see many of his predictions come true. His discoveries led to the wireless telegraph, radio, television, radar and the laser. Yet in 2003, when the *Galloway Gazette*, the local paper for the region in which Maxwell was born, announced a poll to choose the most distinguished resident of all time, it failed even to include Maxwell in its list of famous Gallovidians. When its readers pointed out its error, the *Galloway Gazette* produced a feature describing the great man's life, achievements and legacy, and Maxwell promptly won the poll. Historians of science often argue that if he had lived longer and achieved his full potential then he would be better appreciated and more widely known today. Of course, part of the reason for Maxwell's relative anonymity was the reluctance of his contemporaries to accept his theory of electromagnetism when they could not understand his complex and revolutionary mathematics. He was, simply put, a man who lived several decades ahead of his time.

Chapter 12

INTO THE ATOMIC AGE

Einstein's remark that 'one scientific epoch ended and another began with James Clerk Maxwell' was neither casually made nor hyperbolic. In discovering the electromagnetic nature of light, Maxwell paved the way for the next generations of scientists to make the discoveries which triggered the second scientific revolution and established modern physics. At the heart of their investigations were fundamental questions about matter – in other words, what it was exactly that made, for instance, oxygen different to nitrogen and every other element. In the process of investigating that mystery, we eventually came to understand the primary forces of the universe. The first small step towards that revolution began in 1803, when John Dalton, an English chemist and Quaker, changed the philosophy of chemistry by founding atomic theory.

Born in Cumbria in 1766, Dalton was an extremely bright child who was running his local Quaker school by the age of 12. Unable to attend Oxford or Cambridge because they accepted only Church of England members, he taught himself science sufficiently well to end up teaching natural philosophy and mathematics at Manchester Academy, which was modelled on the Warrington Academy that had flourished under Joseph Priestley. Living a modest and fastidious existence – a neighbour would set his watch daily by the sight of Dalton opening his window to read his thermometer – he published several papers on meteorology, which in turn fostered an interest in the nature of gases.

In 1799, Dalton resigned from Manchester Academy to undertake his own research. Initially, he made his name through his work on gases (Dalton's law describes the pressures within a mixture of gases) and colour blindness, which he diagnosed in himself at a time when the condition was unrecognised. As a consequence, colour blindness is also known as Daltonism.

With a Quaker's propensity for asking fundamental common-sense questions, he wondered where water comes from given that springs continue to run, which led to him devising a simple but effective method for estimating the rate of flow in rivers based on collating meteorological data for England and Wales (one of the most comprehensive data sets obtained until that time). From that data, Dalton obtained average figures for rainfall precipitation, dew precipitation and evaporation loss, which he used to derive the average river flow, reasoning correctly that it would equal the sum of the average rainfall and dew precipitations minus the average evaporation loss.

Dalton's investigation of gases, which he recognised as a collection of tiny particles, led him to consider whether all matter might not be made of similar small particles. The idea of atoms was not new; it went back to the Greeks. But Dalton then became the first to suggest that each element was made of a different and unique atom. By analysing common compounds such as water, carbon dioxide (then called carbonic acid) and carbon monoxide (carbonic oxide) he worked out the relative proportions of each atom in the compounds. From this, he devised a system whereby elements differed from each other primarily in the weight of their individual atoms.

In 1803, Dalton announced his atomic theory. Then in 1808 he laid it out in detail in a book, *New System of Chemical Philosophy*. The main principles of his atomic theory stated that matter could not be divided indefinitely because elements consisted of tiny, indivisible and indestructible particles called atoms; all atoms of a particular element were identical in terms of mass, volume and properties, but atoms of each element were different from the atoms of every other element and had entirely different properties. Compounds (which Dalton called substances) could be turned into one another by breaking the atoms of their different elements apart and recombining the elements in different proportions.

Dalton's single most significant postulation was that atoms of two elements differed from each other only in their mass. This allowed him to become the first person to advance a quantitative atomic theory, and his book had a list of atomic weights. Hydrogen had a mass of one, which was correct. However, his calculations for some elements, such as oxygen and carbon, were incorrect because these tend to combine with other elements in unequal proportions. For instance, Dalton assumed a single oxygen atom combined with a single hydrogen atom to make water, which led him to think oxygen had an atomic weight of eight. Had Dalton known that water contains two hydrogen atoms for every oxygen atom, he would have arrived at the correct atomic weight of oxygen: sixteen.

Although it appeared to make total sense, Dalton's model of the atom was not immediately adopted without resistance. William Wollaston was convinced, but many other scientists found it philosophically challenging, particularly Davy, although his reluctance might have been caused by his notorious jealousy of good ideas put forward by others.

In time, however, Dalton's theories were universally accepted and he was lauded worldwide, receiving honorary degrees and visitations from scientists from all over the world. This attention sat uneasily with Dalton's Quaker beliefs, particularly when it culminated in an audience with William IV. The occasion required that Dalton don scarlet robes, a colour that was forbidden by his religion. Fortunately, Dalton could argue that his colour blindness rendered scarlet invisible to him and he was presented to the king in robes that he insisted were a dull grey.

Increasingly reclusive, Dalton continued to shun fame and glory until the day he died in 1844, when, in complete disregard of his beliefs, a funeral was organised with a procession of 100 carriages and 40,000 people filing past his coffin on public display in Manchester Town Hall.

Dalton's theory remained unchallenged until the discovery of the nature of cathode-ray particles, which indicated that atoms were not entirely indivisible, a discovery that was set in motion when several nineteenth-century scientists asked themselves whether flashes of electric charge that they had seen were the same as Maxwell's electromagnetic waves or maybe some different type of radiation.

One of the scientists to ask that question was William Crookes. The eldest of sixteen children, Crookes was born to a canny

London tailor who through shrewd property investments left a sufficient fortune in his estate for William to give up work and devote his life to science. Crookes set up a laboratory in his London home at which, in 1861, he discovered thallium, a soft metal.

During his investigation of thallium, which he weighed inside a vacuum to minimise atmospheric influences, Crookes became intrigued by the effects of light on substances in vacuums. He developed a method of producing better vacuums (called harder vacuums) in which the air pressure was 75,000 times less than in previous vacuum tubes. Incidentally, this discovery provided the means for Thomas Edison to mass-produce his incandescent light bulbs.

In the 1870s, Crookes used his vacuum technology to produce a tube with a negative electrode (called the cathode) and a minuscule amount of rarefied gas. With this device, thereafter called a Crookes tube, he investigated the radiation emitted by the cathode. At first, these cathode rays appeared to have exactly the same properties as light, leading Crookes to think they could also be wavelike in nature. They travelled in straight lines; they were blocked by solid objects, casting a sharp shadow; and they turned a small silvered wheel in the same way as the heat of sunlight would cause the wheel to turn. All experimental evidence indicated some kind of electromagnetic wave, just like light or radio waves.

Then Crookes discovered he could deflect the cathode rays in a magnetic field. It suggested he was dealing with charged particles, not wavelike radiation with no mass. Baffled by what these charged particles were, Crookes suggested he had discovered a fourth state of matter, but it would fall upon a younger generation of scientists to determine its true nature. In the meantime, Crookes also unwittingly discovered X-rays when he noticed that some photographic plates had become fogged when his Crookes tube was operating, even though they were shielded from light. However, he did not pursue this line of investigation and eventually a German scientist, Conrad Röntgen, got the credit for discovering, identifying and determining the nature of X-ray radiation. And while performing experiments on uranium, Crookes found that he could divide the material into two forms, one of which was much more radioactive than the other. Although he came up with no further evidence, this was the first inkling that radioactive materials change their properties when they emit their radioactivity.

*   *   *

Although Crookes believed cathode rays were not electromagnetic, he had no definitive proof to support his theory. The evidence arrived in 1894, when John Joseph Thomson, one of the few former students of Maxwell who could understand the great man's lectures, discovered that cathode rays travelled considerably slower than light. According to his former mentor's equations governing electromagnetism, the one constant factor was that electromagnetic waves travelled at the speed of light, so the cathode ray had to be something else.

J.J. Thomson was the leading physicist of his generation. Having grown up near Manchester, he had gone to the Owens College, the forerunner of Manchester University, to study engineering. On the death of his father, Thomson's financial position forced him to switch to physics, chemistry and mathematics on a scholarship. At the age of 20 he won a second scholarship, this time to Trinity, Cambridge, where like Maxwell and Lord Kelvin (formerly William Thomson, but no relation) he graduated Second Wrangler in the maths Tripos. He joined the Cavendish Laboratory, becoming Cavendish Professor of Experimental Physics in 1884, although he was so notoriously clumsy that his colleagues often attempted to keep him out of the laboratory.

Physically inept he may have been, but Thomson was an intellectual powerhouse with a gift for devising exactly the right experiment to test a particular hypothesis. In 1897 he turned his skills to examining the single greatest conundrum surrounding cathode rays. Although Crookes had deflected cathode rays in a magnetic field, neither he nor anyone else had managed to repeat the feat in an electric field.

In 1897 Thomson constructed a cathode tube with a harder vacuum than any previous tube. Not only did he deflect the cathode rays, but he also managed to measure the ratio of their charge (e) to their mass (m).

Repeating his experiments with varying amounts of several residual gases in his vacuum tube, he discovered that he always got the same result for the ratio e/m.

Maybe Thomson should have been surprised by this fact, but he was not for the simple reason that he had already guessed that cathode rays were composed of a stream of identical particles. Furthermore, his finding confirmed his suspicion that he was dealing with something much smaller than an atom. Thomson's explanation for this was relatively straightforward: the very large size of the e/m ratio could only be explained if the particle had either a huge charge or a minuscule mass. The relatively small deflection in magnetic and electric fields suggested the latter was more likely.

'The assumption of a state of matter more finely divided than the atom,' Thomson told a lecture audience at the Royal Institution in April 1897, 'is a somewhat startling matter.'

That comment was a considerable understatement. Many of the distinguished scientists in the audience thought he was either wrong or 'pulling their legs', Thomson subsequently wrote. Two years later, he silenced the doubters when he measured the electric charge of a cathode-ray particle by measuring the changes in mass, velocity and charge on oil and water droplets, then using the results and various formulae to derive the charge on a single electron. With e known, Thomson could calculate m. He found the negatively charged mystery particle had a mass more than 1,800 times smaller than the mass of a hydrogen atom, which he said indicated the particles were 'a part of the mass of the atom getting free and becoming detached from the original atom'.

This revolutionary departure from Dalton's model of the atom as the smallest indivisible entity won Thomson the 1906 Nobel Prize in Physics, only the second time in the early history of the prize that it had gone to a Briton. He was knighted in 1908.

* * *

Thomson continued to investigate radiation, but after discovering the electron his greatest contribution to science was his directorship of the Cavendish Laboratory. Under his 35-year leadership, the Cavendish became the world's most renowned centre for sub-atomic research. With an imposing concentration of Nobel laureates in physics and chemistry, many of the world's greatest physicists of the twentieth century trained or worked at the Cavendish, but probably the greatest was a headstrong and rumbustious young New Zealander called Ernest Rutherford.

Rutherford arrived at Cambridge having come second in a competition set up at the time of the 1851 Great Exhibition to enable Commonwealth research students to spend two years studying anywhere in the world. When the winner opted to stay in New Zealand, Rutherford was offered the prize, but said he was only interested if he could use it to go to Cambridge, which until then accepted only research students who had completed their undergraduate training at the university. Rutherford was digging potatoes on the family farm when news that Cambridge had just rescinded the rule reached him. 'That's the last potato I'll dig,' he declared and promptly threw down his spade, postponed his marriage plans and departed for England, arriving in the autumn of 1895.

Radio waves had only very recently been discovered, so Rutherford was assigned to the task of investigating their properties. He swiftly found a way of using radio waves for signalling and for seeing in fog, the principle behind radar, then improved their range until he was sending signals over half a mile, the record at the time. Rutherford's very obvious talent was immediately spotted and he was asked to join J.J. Thomson's research team, incidentally leaving the field of radio waves open for Marconi to pioneer their use for communication.

It did not take long for Rutherford's attitude and dexterity to become highly appreciated by his clumsy supervisor. A rural farming background and scientific training in New Zealand, where science facilities were more limited, had instilled a deeply practical attitude in Rutherford. An advocate of the bits-of-string-and-lumps-of-wax school of investigation, he preferred to make or reuse apparatus, rather than buy it off the shelf, explaining that his training in New Zealand had instilled in him an attitude of 'We don't have the money, so we have to think.'

Tenacious and enthusiastic, Rutherford was relatively unencumbered by the gentlemanly conventions and intellectual constraints of British science. When his work was going well, he would stride around his laboratory good-humouredly singing *Onward Christian Soldiers*. 'We've got a rabbit here from the Antipodes, and he's burrowing mighty deeply,' one of his fellow students famously remarked of him. A young man in a hurry, Rutherford was nicknamed Crocodile because, just as the reptile is unable to see its tail, he always looked forward. His drive to turn conjecture into fact made him one of the original 'demo or die' scientists.

A few weeks before Rutherford joined Thomson's research team, Conrad Röntgen in Germany had discovered X-rays. Thomson assigned Rutherford to investigate their properties. He found X-rays to be a more energetic electromagnetic equivalent of light. He then investigated ultraviolet radiation and the radiation given off by uranium, which had been discovered by Antoine Henri Becquerel in Paris in 1896.

After two years at Cambridge, the terms of the scholarship required Rutherford to leave the university, but he managed to secure a further year. Then with Thomson's help he secured a chair at McGill University in Montreal, where he teamed up with Frederick Soddy, a British chemist. Assisted by Soddy, Rutherford made the first of his great discoveries. In 1899 they found that Becquerel's radiation came in two forms: a short-range variety with low penetrating power (a piece of paper stopped it) and a long-range variety with greater penetrating power. He named them alpha and beta radiation. The next year he discovered a third form of radiation that was not deflected by a magnetic field and which could penetrate most materials. Rutherford called it gamma radiation.

By analysing the products of radioactive decay, Rutherford and Soddy discovered that when an element emitted alpha or beta radiation it turned into a new element. It meant radioactivity was an atomic phenomenon (strictly speaking, the *nucleus* emitted alpha and beta particles, so it was a *nuclear* phenomenon, but the discoveries that clarified those details lay in the future). He also found that the intensity of the radiation was constant for a given quantity of a radioactive substance and that it declined at a constant rate. Rutherford called the time taken for a radioactive material to lose half its radioactive intensity its half-life.

These discoveries overturned the popularly accepted classical belief that elements were immutable – 'atom' is a Greek construction meaning unsplittable – and gained Rutherford the 1908 Nobel Prize in Chemistry. Although thrilled to receive the award, Rutherford was less impressed by the classification, saying that he had observed many rapid transformations among radioelements, but none so rapid as his transformation from a physicist into a chemist. Remarking that 'All science is either physics or stamp collecting,' he regarded chemistry as an inferior science.

By the time he was awarded the Nobel Prize, Rutherford was back in England, where in 1907 he took over as professor of physics at Manchester University from Arthur Schuster, who had declared he would relinquish his chair only if it was given to Rutherford.

Rutherford acquired some radium from Vienna. Working with Hans Geiger, with whom he developed an ionisation chamber that Geiger later refined to create his well-known Geiger counter, Rutherford turned his attention to alpha rays, which he now discovered were not rays at all but particles. More specifically, they were analagous to helium atoms that had lost two negatively charged electrons and were therefore deflected strongly in a magnetic field because they had an excess two units of positive charge.

In 1909, Rutherford achieved his greatest breakthrough, an achievement that many regard as defining the moment the modern age started.

At that time, the atom was thought to be like a piece of fruitcake: a positively charged mass (like the sponge) dotted with negatively charged electrons (like the sultanas).

Firing a beam of alpha particles at gold foil only one fifty-millionth of an inch thick, Rutherford found that most of the particles went straight through the foil. However, a very small number were deflected, in some instances by ninety degrees. An even smaller amount of the alpha particles bounced straight back. 'It was as if you fired a 15-inch shell at a sheet of tissue paper and it came back to hit you,' Rutherford said afterwards.

The implications were clear. They made nonsense of the sponge cake model of atoms.

Given that the foil was more than 2,000 atoms thick and most of the particles had passed straight through it, the only viable explanation was that atoms were mostly empty space. However, the deflection of a small proportion of the alpha particles, occasionally very markedly, indicated that atoms had a tiny highly concentrated area of mass and positive charge. Rutherford called this area the nucleus, which he calculated was about 100,000 times smaller than the entire atom. Or, as it is sometimes described in the well-known analogy: like a grain of sand at the centre of the Albert Hall.

In 1911, Rutherford announced his model of the nuclear structure of the atom: a positive nucleus surrounded by a cloud of tiny negative electrons, their mass far too small to deflect a beam of alpha particles. This second great discovery was of even greater significance and implications than his discovery of alpha, beta and gamma radiation, and by rights he should have received a second Nobel Prize, but for some reason it never came, possibly because many scientists struggled with the concept that every atom was almost entirely made up of empty space.

Rutherford's model was initially rejected. However, his reputation and achievements attracted leading physicists from all over the world to Manchester, among them a young Danish football fanatic called Nils Bohr. A year after arriving in 1912, Bohr elaborated on Rutherford's model to produce his quantum model of the atom in which the negative electrons orbited the positive nucleus at fixed distances and therefore at fixed energy levels. The discovery won Bohr (and the Cavendish yet again) a Nobel Prize in Physics.

Bohr's discovery was not enough on its own to persuade sceptics of Rutherford's nuclear model of the atom, but a discovery in the same year by Henry Moseley, a protégé of Rutherford with the reputation of being the youngest and most brilliant of the gifted young men he had recruited to his research team, did the trick.

Moseley bombarded crystals with X-rays. When he examined the wavelengths of X-ray radiation emitted by the crystals, he found the wavelengths decreased in a very regular way with the increasing positive nuclear charge of the atoms in the crystals. This indicated to Moseley that nuclear charge had a greater influence on an element's properties than any other characteristic. Moseley then rearranged the periodic table in order of nuclear charge (later called atomic number) and found that his new table overcame all the faults and restrictions of the previous version, which had been drawn up by the Russian chemist Dmitri Mendeleev in 1869.

Moseley's discovery was of fundamental importance in establishing Rutherford's model of nuclear structure and understanding its implications. It meant that an element's main properties are determined by a whole number defining its nuclear charge. Moseley also predicted that four missing elements between aluminium (atomic number 13) and gold (79) would be discovered. He was correct, although by the time that the four elements had been uncovered the First World War had broken out and Moseley had enlisted with the Royal Engineers. By August 1915 he was dead, shot through the head by a sniper during the ill-fated assault on Gallipoli.

If Moseley had not died in battle there is every possibility he too would have won a Nobel Prize (it is not awarded posthumously). He would have gone on to do further work and, judging by the talent he showed at such a young age, would have become one of Britain's best-known and greatest scientists. The tragedy of his death at 27 brings to mind Newton's comments about Roger Coates, his collaborator on *Principia*: 'If he had lived, we would have known something.' Within the scientific community, Moseley's achievements are recognised for the vital evidence they provided for Rutherford's model of nuclear structure. 'The Rutherford work was not taken seriously,' Niels Bohr said in 1962. 'We cannot understand today, but it was not taken seriously at all. There was no mention of it any place. The great change came from Moseley.'

While Moseley and many of the best scientists of his generation were at war, Rutherford conducted military research for the government; research that led to the development of sonar. Then in 1917, working alone because all his collaborators were on the battlefields of Europe, he started the work that led to his last great discovery. With typical understatement, Rutherford described his work at this time as 'playing with marbles' and in essence that is exactly what he was doing, firing one type of marble (an alpha particle) at a collection of another type of marble (atoms in a gas) to see what happened. However, in a letter to Bohr, he divulged his true intention: 'I am also trying to break up the atom by this method ... regard this as private.'

As usual, Rutherford used an apparatus that was so simple it bordered on being crude. He built a sealable brass cylinder that he could fill with a gas. The tube had a glass phial at one end into which Rutherford could drop a source of alpha particles. The alpha particles would then pass through the glass and interact with the gas. If the interaction generated subatomic particles they would be picked up at the other end of the tube by a crude device for measuring radiation called a scintillation counter, in this instance a zinc sulphide screen that flashed every time a subatomic particle collided with it.

With hydrogen in the tube, particularly bright scintillation flashes were seen; these were single protons liberated from the nuclei of the hydrogen atoms. With oxygen, the number of flashes dropped off because the oxygen atoms absorbed some of the alpha particles. But with nitrogen, although the number of flashes characteristic of alpha particles declined again, there were also occasional bright flashes characteristic of protons, or, as they are more commonly known, the nuclei of hydrogen atoms. This was an astonishing result. Before he had introduced the alpha particle source, the tube had contained only nitrogen. Now it also apparently contained hydrogen. Somehow he had turned one element into another – the ultimate ambition of the alchemists.

Rutherford waited until he published his results in 1919 before making a public announcement on his momentous achievement. When it came in the fourth part of his paper, Rutherford famously underplayed his triumph: 'From the results so far obtained it is difficult to avoid the conclusion that the long-range atoms arising from collision of alpha particles with nitrogen are not nitrogen atoms but probably atoms of hydrogen ... If this be the case, we must conclude that the nitrogen atom is disintegrated.'

The scientific community, the media and the rest of the world were not quite so reserved. Rutherford had split the atom, they trumpeted.

If it seems out of character for Rutherford to have downplayed his achievement, the explanation can probably be found in the fact that he was meant to be engaged in anti-submarine research at that time. Immediately after announcing his discovery, he was called in front of a military overseers' committee. Asked to defend his actions, Rutherford was quietly defiant: 'If, as I have reason to believe, I have disintegrated the nucleus of the atom, this is of greater significance than the war.'

But had Rutherford really split the atom? Some historians argue that it was his protégés, John Cockcroft and Ernest Walton, who were the first atom splitters. Or perhaps the honour should go to a group of Americans led by Enrico Fermi, an Italian-born physicist. He initiated the first controlled, self-sustaining nuclear chain reaction in a nuclear pile constructed in a squash court beneath the West Stands of Stagg Field, the University of Chicago's athletic stadium, as part of the US Army's Manhattan Project to build an atomic bomb.

Meanwhile, pedants point out that *atoms* are routinely split because all chemical reactions, even the dissolving of common salt in water to make brine, involve the transfer of electrons from one atom to another. To satisfy pedants, the question should be: did Rutherford really split the *nucleus*? The answer is: yes, Rutherford was undoubtedly the first to knowingly create a nuclear fission reaction.

When Rutherford bombarded the seven-proton nitrogen nucleus with the two-proton helium nucleus of an alpha particle, the fast and heavy alpha particle chipped a single-proton hydrogen nucleus off the nitrogen nucleus and attached itself to the nitrogen nucleus to make an eight-proton oxygen nucleus. Or, as Rutherford described it: 'We must conclude that the nitrogen atom is disintegrated under the intense forces developed in a close collision with a swift alpha particle, and that the hydrogen atom which is liberated formed a constituent part of the nitrogen nucleus.'

Rutherford is often described as the most accomplished experimenter since Faraday. Although the accolade is fair, it does Rutherford a slight injustice. Of his three great discoveries (the nature of alpha, beta and gamma radiation in 1903, the nuclear structure of the atom in 1909, and the splitting of the nucleus in 1917), only the third was the result of experimentation. The other two discoveries were entirely theoretical concepts backed by mathematical analysis and experimental evidence, but because of his experimental genius the significance of Rutherford's great theoretical achievements is frequently overlooked.

In the case of his splitting of the nucleus, Rutherford had in a sense finally achieved what the alchemists of the Middle Ages had always wanted. He had transmuted one element into another. Einstein, who published his theories in the same period that Rutherford was investigating radiation and atomic structure, declared his contemporary to be 'a second Newton', the man who 'tunnelled into the very material of God'.

However, Rutherford's nuclear reaction was not particularly efficient. Only about one alpha particle in every 300,000 collided with a nitrogen nucleus to create oxygen and knock off a proton, so only a minuscule proportion of the nitrogen nuclei were split – it was almost as if Rutherford had split the nucleus incidentally. A totally intentional and efficient nuclear reaction in which the nuclei of all the atoms were split was achieved in April 1932, when Cockcroft and Walton bombarded lithium atoms with protons to make helium atoms. Like many previous and subsequent crucial events in the history of nuclear physics, it happened at the Cavendish Laboratory in Cambridge, and, only a fortnight later, James Chadwick, also at the Cavendish, discovered the neutron. For nuclear physics and the Cavendish Laboratory, 1932 was yet another *annus mirabilis.*

* * *

All these discoveries were overseen by Rutherford, who in 1919 had become the director of the Cavendish Laboratory, bringing Chadwick with him from Manchester as his deputy and closest collaborator.

Born in Cheshire, Chadwick was a bright but extremely reserved youngster with a gift for mathematics. At the age of 16, he applied for a scholarship to read mathematics at Manchester University. When he turned up for his interview, Chadwick was mistakenly interviewed by Rutherford as a potential physics student. Too shy to point out the mistake (in later life he feared public speaking so much that he'd sometimes be physically ill at the prospect of it), he answered every question. By the time he'd finished, he'd been offered a place to read physics and was so impressed by Rutherford that he decided to accept it.

Chadwick struggled with the noise and size of first-year classes but found his feet in his second year, when Rutherford realised his talent and nurtured him. Rutherford set Chadwick a third-year research project involving comparisons of different radium sources, but left a deliberate mistake in the method to test Chadwick. Although Chadwick noticed the mistake, he didn't dare correct his professor. Instead he ignored it, leading Rutherford to think Chadwick had missed it, even though his project was good enough to garner a first-class honours degree and a joint publication with Rutherford.

Chadwick stayed at Manchester, working on research that led to Rutherford's discovery of the nucleus, his own Masters degree and a recommendation by Rutherford for the same 1851 Exhibition scholarship that had brought Rutherford to England eighteen years earlier. Required by the scholarship to travel, Chadwick opted to move to Berlin in 1913 to work with Geiger on radioactivity. His research was going well and he had met leading German scientists, including Albert Einstein, when the First World War brought a sudden end to his work. Geiger was called away as a reserve officer and Chadwick was interned as an enemy alien at Ruhleben racecourse in the Berlin suburb of Spandau. Housed with five other internees in a stable built to hold two horses, he suffered inadequate food and bitter cold. But he continued his research in captivity, establishing a science society of which he was secretary and using toothpaste as a radioactive source in his experiments.

In 1918, when the war ended, Chadwick returned to England. Four years of imprisonment had matured him and made him confident of his capabilities, but they had also left him impoverished and with his health shot. Rutherford offered him a job at Manchester, then took him on to Cambridge in 1919, when he was appointed the Cavendish Professor. As Assistant Director of Research, Chadwick became Rutherford's closest colleague, overseeing the laboratory's researchers by day and revising Rutherford's book on radioactivity by night in a house that was so cold and draughty that he did the work wrapped up in an overcoat and wearing gloves.

By the early 1920s, Rutherford and Chadwick were convinced that the disparity between nuclear mass and charge could only be explained by the existence of neutrally charged particles, which Rutherford called the neutron. They split the nuclei of several of the lighter elements in the same way that Rutherford had earlier split nitrogen, but they found that as the elements became heavier the increasing positive charge on the nuclei repelled the alpha particles and they were unable to split heavier atoms. Somehow they needed to find a way of splitting larger, heavier atoms. The answers came from a new generation of nuclear scientists supervised by Rutherford at the Cavendish.

The first breakthrough came in a discovery by Chadwick after discussions with Rutherford. The two men had questioned why the mass of the nuclei of successive elements in the periodic table increased at a faster rate than their positive charge. For instance, the hydrogen nucleus has a single charge corresponding to the single proton that creates its mass. However, the helium nucleus has a mass equal to four protons but a charge equivalent to only two protons. This imbalance begged a further question of why the increasing numbers of protons in the nucleus didn't repel each other and rip the atom apart.

Several explanations were suggested, including that the nucleus contained electrons that neutralised the charge of the protons but didn't affect its mass. Complex theoretical reasons, however, ruled out this explanation. Then, after more than a decade of experimentation, Chadwick finally found the answer after reading in January 1932 of work by a French husband and wife team, Irène Joliot-Curie, daughter of Marie Curie, and Frédéric Joliot. Having bombarded beryllium with alpha particles, they had noticed that some kind of invisible radiation was emitted, which they couldn't identify but which was sufficiently energetic to knock protons out of paraffin. They speculated that it might be a form of gamma radiation, but Chadwick instantly guessed what they had seen: a particle with the same mass as a proton but without the proton's positive charge, which made it particularly penetrating, impervious to magnetic and electric fields and therefore very difficult to detect. It was Rutherford's elusive neutron.

For three weeks Chadwick worked night and day. Then in February 1932 he announced his discovery in a letter to the scientific journal *Nature* titled 'Possible existence of a neutron'. The discovery of the third and final part in the jigsaw of atomic structure changed not only the field of particle physics but the course of history. By far the most useful particle for initiating chain reactions, the proton was soon being used in particle accelerators. Knowledge of neutrons and their behaviour was also critical in the development of the nuclear bomb, the first two of which were dropped just ten years after Chadwick's discovery won him the 1935 Nobel Prize in Physics.

INTO THE ATOMIC AGE

When the Second World War broke out in September 1939, history came close to repeating itself. Chadwick – who was on a fishing holiday in Sweden – found himself unable to cross enemy lines to return home. Fortunately he obtained a flight from Stockholm to Amsterdam, from where a few days later he was able to return to Liverpool to resume his work there as professor of physics. By then, the apocalyptic destructive potential of nuclear chain reactions had been realised in theory and Chadwick was asked by the government if a nuclear bomb was possible in practice.

A year later, after conducting experiments at Liverpool, Chadwick had the answer. Not only was a nuclear weapon possible, but it did not need the ton of uranium that American scientists predicted would be required to produce a nuclear chain reaction. Chadwick calculated that a weapon would need only a few kilograms of uranium. With several other leading British physicists, he formed a group to produce a British nuclear bomb by 1943, but after America entered the war in December 1941 Chadwick petitioned for the two allied nations to cooperate on developing it. In August 1943 the Quebec Agreement on Anglo-American collaboration was signed by Churchill and Roosevelt. British scientists then moved to America under the leadership of Chadwick to work on the Manhattan Project. In August 1945, the fruit of their endeavours, the atomic bombs Little Boy and Fat Man, were dropped on Hiroshima and Nagasaki with devastating effect.

* * *

Among the other British scientists working with Chadwick and the Americans on the Manhattan Project was a remarkably versatile engineer, mathematician and physicist called John Cockcroft, who split the atom to perform the first artificial nuclear transformation only weeks after Chadwick discovered the neutron.

Born in 1897 to a family of cotton manufacturers on the Yorkshire–Lancashire border, Cockcroft began his university career studying mathematics at Manchester, but the First World War intervened. After serving with the Royal Field Artillery he gained an interest in electrical engineering and in 1918 he elected to study the subject at Manchester College of Technology instead of returning to mathematics. After graduating, he served a two-year apprenticeship at a Manchester electrical engineering factory, then took the mathematics Tripos at Cambridge in 1924.

When he graduated, his rare combination of practical engineering and theoretical mathematics skills were seized upon by Rutherford, who recruited Cockcroft to join his band of experimentalists at the Cavendish Laboratory.

By the mid-1920s, Rutherford and Chadwick were starting to realise that their method of bombarding elements with alpha particles to split their nuclei was running out of steam as the elements became heavier and their nuclei repelled the positively charged particles. In 1928, George Gamow, an ebullient Russian émigré physicist attracted to the Cavendish by Rutherford's reputation, published a paper showing how the new field of quantum mechanics could provide proof of how alpha particles could tunnel through nuclear barriers to penetrate the nucleus. As one of the Cavendish's few theoreticians, Cockcroft immediately spotted the potential implications of Gamow's theory.

Working with a young Irish student, Ernest Walton, Cockcroft built a particle accelerator with a voltage multiplier. It developed a charge of 710,000 volts to accelerate protons to a state of sufficiently high energy to penetrate the nucleus of heavier elements.

Begging and borrowing components from Cockcroft's contacts in the engineering industry, they slowly assembled their particle accelerator. By 1932, Rutherford was losing patience and demanded some immediate results. Then, on 14 April 1932, Cockcroft and Walton satisfied Rutherford's demand in spades. Closeting themselves in a small lead-lined wooden hut beside the accelerator, and with a zinc sulphide particle detection screen tacked to the hut's wall being their only concession to health and safety, they bombarded a sample of lithium with high-energy alpha particles. When the single proton of each alpha particle collided with the three-proton lithium nucleus, it split the nucleus to create two two-proton helium nuclei.

Unlike Rutherford's splitting of the nitrogen nucleus, this was an efficient reaction. Provided that the alpha particles had sufficient momentum, they would split every lithium nucleus with which they collided. The discovery was as important as any other in nuclear physics, but Cockcroft and Walton had to wait until

1951 to share the Nobel Prize in Physics for the 'transmutation of atomic nuclei by artificially accelerated particles'. With so many vital discoveries in particle physics underpinning the second scientific revolution of the late nineteenth and early twentieth century, the Nobel Institute felt it should first honour other giants of the field before coming to Cockcroft and Walton.

* * *

Several other scientists who worked at the Cavendish or who studied under Rutherford made discoveries that threw further light on the dark mysteries of atomic structure and the nucleus. Among them was Patrick Blackett, who as Professor of Physics at Birkbeck College of the University of London, was the first to provide experimental proof of Einstein's prediction in his special theory of relativity that matter may be created from energy according to his iconic equation $E=mc^2$.

In 1935, while passing gamma rays through lead, Blackett discovered that some of the rays disappeared. Energy had been converted into matter, creating an electron and a positron in this case. The discovery of a positron, the positively charged antimatter counterpart of the electron, was highly significant. Its existence, predicted by Paul Dirac in 1928, neatly tied up some of the unanswered questions concerning nuclear structure.

One of the giants of twentieth-century physics, Dirac is considered by Stephen Hawking to have been the greatest physicist since Newton. Dirac's work finally answered some of the perplexing conundrums thrown up by Rutherford's model of the atom. Although elegant and useful for explaining many chemical processes, the Rutherford model had prompted two elementary questions: why didn't the negatively charged electron cloud collapse inwards onto the positively charged nucleus? And why didn't the concentration of positively charged protons in the nucleus repel each other and blast the nucleus apart? The answer to these questions is highly complex and involves mathematical theories beyond the scope of this book, but to simplify the science greatly: quantum mechanics prevents the electron cloud from imploding into the nucleus; and a strong force, more powerful over distances of a ten-thousand-billionth of a metre than the repulsive forces between protons, binds the nucleus together.

The first conundrum – why the electron cloud doesn't collapse into the nucleus – was solved by Niels Bohr, the football-obsessed Dane attracted to Manchester in 1912 by his great friend Rutherford.

Bohr's great genius was his ability to draw upon different theoretical views of the physical universe and to knit them together into new theories that stood up to mathematical analysis and experimental testing. In the case of his atomic model, he drew upon quantum theory formulated by several scientists but primarily based on discoveries by Albert Einstein and Max Planck in Germany. Bohr explained that electrons could occupy only fixed orbits around the nucleus in a way that was similar to the process of climbing a ladder being possible only by standing on the rungs, not on the spaces between the rungs. However, many scientists found this interpretation difficult and shocking, reasoning that if it was applied to more familiar entities, it would mean that a car accelerates from, say, 0 to 60 m.p.h. in a series of jumps, intermediate speeds never even existing. However, Bohr had good reasons for his quantum model of the atom, mainly resting in the discovery that electrons emit or absorb energy in the form of discrete packets of energy called quanta. Whenever they absorb or emit these quanta, electrons jump into an orbit further or nearer to the nucleus. Each orbit can contain only a fixed maximum number of electrons – again for complex mathematical reasons related to the way in which electrons can be both particles and waveforms simultaneously – so the electrons do not all jump into the orbit nearest the nucleus.

By the mid-1920s, the physical world had been described by two great physics theories: quantum mechanics, which thanks to Bohr and others described the behaviour of all the very small things in the world, such as atoms and electrons, and relativity, which described the strange ways the world behaves when travelling at or near the speed of light.

Dirac's great achievement was to bring the two theories together through a wave equation for the electron that took into account the special theory of relativity. Although Dirac casually referred to his work as 'just playing with equations and seeing what they give', his discoveries provided vital evidence to help explain why the concentration of positive charges in atomic nuclei doesn't blow them apart.

Dirac believed fervently that the mathematical simplicity and elegance of an equation is a measure of its accuracy. 'A theory with mathematical beauty is more likely to be correct than an ugly one that fits some experimental data,' he said. 'God is a mathematician of the very highest order and he used very advanced mathematics in constructing the universe.'

For all its supposed divine provenance, to most of us the equation means absolutely nothing. But when Dirac unveiled it to the world's leading physicists and mathematicians, they were literally stunned by its beauty:

$$\left(\beta mc^2 + \sum_{k=1}^{3} \alpha_k p_k c \right) \psi(\mathbf{x},t) = i\hbar \frac{\delta \psi(\mathbf{x},t)}{\delta t}$$

Published in 1927, it was the height of elegance to anyone who understood it and the definitive statement on the electron. It explained properties that were entirely new and predicted other properties that hadn't yet been discovered. In Einstein's words, it was 'the most logically perfect presentation of quantum mechanics'.

However, the most startling and significant prediction came from the equation's answer, or rather its two simultaneous answers. One of the solutions was positive, while the other was the same value, but negative – such as, for instance, +1 and −1.

After much head-scratching and several attempts to make the solution fit the Rutherford–Bohr model of the atom, Dirac realised four years after he published his equation that it was predicting an entirely new particle. With the same mass as an electron, but a positive charge equal to the electron's negative charge, it was a positive twin to the electron – an anti-electron, or as we now know it a particle of antimatter.

Not only did the equation predict antimatter equivalents of all subatomic particles – the proton, neutron and electron – but it also made it clear that none of them could exist in isolation. Only as positive–negative pairs would all the quantum properties cancel each other out, leaving only the particles' mass. And if an electron encountered a positron, Dirac predicted that the two charges would cancel and the pair would be annihilated. The combined mass would transform into radiation in the most dramatic expression of Einstein's celebrated equation $E=mc^2$.

Two years after Dirac publicised his antimatter explanation for his equation, Carl Anderson, an American physicist working at the California Institute of Technology, was working on cosmic rays. Looking through some photographs of particle tracks in a magnetic field, Anderson noticed one that looked exactly like the particle of an electron except it curved the opposite way. Anderson immediately recognised it as Dirac's predicted positive electron and named it the positron. Another two years later, in 1935, Blackett provided the first experimental proof of the positron's existence. And a further twenty years later, the Italian-American physicist Emilio Segrè discovered the anti-proton.

All posited types of antimatter particle have now been discovered – some scientists even predict entire antimatter universes, although evidence of the existence of these has never been found – and the predictions made in Dirac's equation have been found to be correct in every respect.

> Dirac isn't exactly a household name, but I think he should be. And Dirac may not sound British, but British he was – and a physicist's physicist.
>
> Obsessed by beauty, simplicity and mathematics, Dirac once famously said that God used beautiful mathematics in creating the world, although when he said God, he probably meant nature. He was a confirmed atheist.
>
> Kathy Sykes

Widely regarded by top-rank physicists as among the greatest physicists since Newton, Dirac's genius is difficult for most of us to appreciate. Anyone who does not understand advanced mathematics would struggle to understand why Dirac is so special. His equation is beautiful, but only the mathematically literate can share its full beauty. The conundrum is possibly best explained by the great American physicist and father of quantum electrodynamics, Richard Feynman, who developed Dirac's work to win a Nobel Prize of his own.

Feynman said: 'If you are interested in the ultimate character of the physical world, or the complete world – and at the present time our only way to understand that is through a mathematical type of reasoning – then I don't think a person can fully appreciate, or in fact can appreciate much of, these particular aspects of the world – the great depth of character of the universality of the laws, the relationships of things – without an understanding of mathematics.

'Don't misunderstand me, there are many, many aspects of the world that mathematics is unnecessary for, such as love, which are delightful and wonderful to appreciate and to feel awed and mysterious about. And I don't mean to say that the only thing in the world is physics, but you were talking about physics and if that's what you're talking about, then to not know mathematics is a severe limitation in understanding the world.'

The equation won Dirac a Nobel Prize in Physics 'for the discovery of new productive forms of atomic theory', which initially he did not want because he hated publicity. But Rutherford persuaded Dirac that the publicity would be much worse if he turned the prize down.

Dirac was certainly a most unusual man, possibly autistic, who blamed his emotional frailties on his father (who had bullied his wife and children, insisting they spoke only French at home even though they lived in Bristol). Dirac once said, 'I never knew love or affection as a child,' and there are many stories and anecdotes that illustrate his strangeness. On a cruise ship to Japan in 1929, his companion, the great physicist and unabashed hedonist Werner Heisenberg, was enjoying himself flirting and dancing with women on the ship. Fusty and geeky, Dirac recoiled from any socialising or small talk.

'Why do you dance?' he asked his more outgoing friend.

'When there are nice girls, it is a pleasure,' replied Heisenberg.

Dirac thought about this and then said: 'But how do you know beforehand that the girls are nice?'

Another story relates what happened when Dirac's wife asked what he would do if she left him. Again Dirac thought about it. Then he said, 'I would say: "Goodbye, dear."'

Similarly, Dirac criticised the physicist J. Robert Oppenheimer's interest in poetry, saying, 'The aim of science is to make difficult things understandable in a simpler way; the aim of poetry is to state simple things in an incomprehensible way. The two are incompatible.'

Niels Bohr was a good friend of Dirac and described him as having the 'purest soul' of all physicists and also as not having a single trivial bone is his body. On one occasion, Bohr did not know how to finish a sentence in a scientific article he was writing. Dirac did not hesitate in replying, 'I was taught at school never to start a sentence without knowing the end of it.'

Some people have suggested that Dirac's reticence, literal-mindedness, rigid patterns of behaviour and self-centredness indicate he was truly autistic. If he was, then perhaps it provided him with the characteristics of concentration, obsessiveness and determination that he brought to his mathematics and physics. Others who knew Dirac well, however, claim it simply isn't true to suggest that he had no interest in ordinary things, such as books and films. He was inspired by Beethoven and Rembrandt, but he also loved Mickey Mouse films, for example, and in later life was a great fan of Cher, but he didn't talk about them.

Chapter 13

THE EXPEDIENCY OF WAR

War has a remarkable way of concentrating the mind, not least in the field of science. From the late 1930s until 1945, all efforts and resources focused on the single goal of defending the nation. As a consequence, scientific discovery in the Second World War went into overdrive. But this was not a time for blue-sky science or investigations of the nature of our existence; all scientific endeavour was aimed at fostering discoveries and developing technologies that would ensure our survival and the protection of our way of life. Interestingly, the key British discoveries and innovations of the Second World War – penicillin, computing and the jet engine – all had their beginnings in work that was abandoned, neglected or overlooked in the pre-war era, only to be seized upon when budgetary constraints and intellectual qualms were pushed aside so that science could save the nation.

Probably the best-known breakthrough of the war, and certainly one of the most important advances ever made in medicine, began in 1928 with a Monday morning clean-up in a hospital laboratory in central London.

The son of a Scottish sheep farmer, Alexander Fleming was a young research scientist with a fascination for bacteria that had brought him a profitable side practice treating the syphilis infections of prominent London artists. Already well known from his earlier work, which included the discovery in 1922 that tears and mucus were given antibacterial properties by a protein called lysozyme, by the summer of 1928 Professor Fleming had developed a reputation as a brilliant researcher, but a quite careless lab technician. He often forgot cultures that he worked on, and his lab in general was usually in chaos.

Before leaving his laboratory for his annual summer leave, Fleming had left on his lab bench some culture plates smeared with *Staphylococcus* bacteria, the germ that causes septic infections. While Fleming relaxed on his holiday, serendipity struck. A spore of a rare variety of mould drifted into Fleming's lab from the mycology laboratory one storey below. Had Fleming been a more careful technician, he might have covered his *Staphylococcus* cultures and placed them in an incubator. But he hadn't. And so the spore landed on Fleming's open petri dishes. The next stroke of good fortune was that London was hit by a cool spell. Below average temperatures for August that year provided the ideal conditions for the mould to grow, spreading like a fluffy lawn across the surface of the dishes.

On 3 September 1928, the final events in Fleming's serendipitous sequence occurred. As Fleming described it, 'I certainly didn't plan to revolutionise all medicine by discovering the world's first antibiotic, or bacteria killer. But I guess that was exactly what I did.'

Arriving in his lab at St Mary's Hospital Medical School, most probably sporting his usual bow tie, Fleming noticed that many of the culture dishes he had left out before his holiday were contaminated with mould. He threw the dishes into a basin of disinfectant and was about to get on with his day's work when a visitor arrived in the lab. Had the visitor not arrived, then Fleming most probably would never have discovered penicillin's antibiotic properties. However, he needed to show his visitor what he had been researching, so he retrieved some of the submerged dishes that he had just discarded.

Most scientists dismiss the idea of Eureka moments. Discoveries and insights, they say, are the result of long processes of careful investigation and iterative improvements to their experimental methods. But not Fleming. As he showed his petri dishes to the visitor, he noticed a clear halo around the invading yellow-green mould growth and was struck by a sudden insight.

'That's funny,' said Fleming to his visitor, pointing to the strange ring.

Fleming's long-term interest in ways of killing bacteria responsible for a vast range of diseases immediately alerted him to what had happened. It appeared that within the halo around the mould the *staphylococcus* bacteria had died and no new growth had invaded the area.

In fact, Fleming was not the first to recognise this particular mould's antibacterial properties. In 1875, John Tyndall, an Irish physicist who succeeded Faraday at the Royal Institution, spotted its effect but he had failed to recognise its potential uses. However, Fleming's background as a bacteriologist alerted him to the significance of his discovery. He made a careful drawing of what he saw and then set about trying to find out more about the mystery substance.

Fleming cultured the yellow-green growth, then filtered off a 'juice' from a felt-like layer, which he identified as *Penicillium notatum*, a mould similar to that which grows on stale bread, so he named the juice penicillin. This substance, he discovered, killed some other bacteria apart from staphylococcus, but not all: some bacteria grew well in its presence.

In further experiments, Fleming discovered there were no harmful side effects when he injected penicillin into rabbits and mice; however, he failed to take the vital next step of injecting it into an infected mouse to find out if it also eradicated infection. One of Fleming's assistants cultivated some *Penicillium* in milk, then drank it without ill effect. However, Fleming's most significant discovery was that penicillin was harmless to white blood cells at concentrations that were sufficient to kill bacteria. After all, there were plenty of substances that could kill bacteria – bleach, for instance – but these were poisonous to most other cells, including human ones.

Fleming's discovery of penicillin changed the course of history. It was where the era of antibiotics began.

So what was the genius of Alexander Fleming? Partly it is as Louis Pasteur said, 'Chance favours the prepared mind.'

Fleming's mind had been prepared in the First World War, when he'd seen people dying of infections. Then, during the 1920s, Fleming had been thinking and experimenting with ways that might inhibit the growth of bacteria. And when chance walked that mould spore on to that plate, Fleming's mind was already prepared to recognise what had happened, to think about it and to understand it. And to see where it might lead.

> When Fleming couldn't work out how to produce enough of it to test on animals, let alone humans, he was generous enough to send the penicillin mould around to different labs around the world. That's really important in science because scientists who keep discoveries to themselves never get anywhere.
>
> Paul Nurse

Eventually Fleming ran out of steam. Not being a chemist, he could not isolate or identify the exact active agent in his mould and he lacked knowledge of a way to grow *Penicillium* reliably and in large quantities. And without knowledge of its structure, he had no idea of how he might synthesise it.

Fleming lost faith in his discovery. Penicillin's action appeared to be slow, it was difficult to produce in quantity and it seemed unlikely to last long enough in the human body to fight infection. Convinced there was little future in penicillin, Fleming abandoned his research. His final act was to preserve his unusual strain of *Penicillium notatum* in case it might be useful at a later date. He then published his results in 1929 in the *British Journal of Experimental Pathology*. Little attention was paid to his article.

Throughout the 1930s, Fleming occasionally returned to his penicillin work and he frequently encouraged his research students and other scientists to seek ways to mass-produce and refine usable penicillin, but nothing came of it.

A decade after Fleming's initial discovery, two scientists at Oxford University got their hands on a small sample of his mould. Howard Florey was an energetic, driven Australian-born British pathologist and Ernst Chain was a German-born Jewish biochemist who had escaped Nazi Germany when Hitler came to power. They had been working on lysozyme, the anti-bacterial enzyme in tears, mucus and saliva discovered by Fleming, when they came across Fleming's notes on the juice from his *Penicillium* work.

Recruiting a team that included the talented biochemist and great experimentalist Norman Heatley, they immediately set to work. Chain, a particularly accomplished biochemist, cultivated Fleming's strain of *Penicillium*, then found a way of extracting and purifying penicillin without destroying its antibacterial effect. A scientific *tour de force*, the process was extraordinarily difficult;

after eighteen months they had collected only one-tenth of a gram of the yellow powder, but it was enough to begin a series of experiments.

Chain and Florey discovered that just one-thousandth of a gram of their powder, heavily diluted, was still sufficiently potent to combat lethal *streptococci* infections. An hour after injecting eight mice with virulent *streptococci*, they gave four of them penicillin injections. Within a few further hours, all four untreated mice were dead; the next day, all the mice treated with penicillin were alive and thriving. If it worked in mice, they reckoned it would probably work in humans.

'It looks promising,' Florey remarked, typically laconic in his assessment of one of the most significant experiments in medical history.

In tandem with Chain, Florey also discovered that penicillin acted by blocking normal processes of cell division in bacteria, rather than behaving like antibacterial enzymes or antiseptics, which were much more toxic to human and animal cells. Chain then correctly deduced its structure.

In 1940, Florey and Chain published their results in *The Lancet* to widespread acclaim, including from Fleming. When Florey told Chain that Fleming had contacted him and was going to visit their laboratory, Chain was shocked.

'Good God!' he said. 'I thought he was dead.'

However, Florey and Chain were distracted by more pressing concerns. At that time, a German invasion of Britain seemed imminent. Worried that their work might be disrupted or forever lost, the two men smeared spores of their *Penicillium* culture on the linings of their coats so that they might be able to resume their research whenever time and circumstances allowed it. Fortunately the German invasion never happened, so Florey and Chain continued their work, largely ignored by the military authorities and pharmaceutical companies, whose attention was then focused entirely on producing vaccines for the war effort.

With perseverance, ingenuity and determination in the face of disinterest from the government and industry, Florey and Chain turned their lab into a small penicillin factory, waiting for the right time to test their creation on humans. Humans are 3,000 times heavier than mice, so they needed at least 3,000 times more penicillin than they had used in their experiments on mice. They grew it in bed pans.

The long-awaited opportunity came in 1941, when a doctor at Oxford's Radcliffe Infirmary happened to hear about their work. Dr Charles Fletcher had a patient named Albert Alexander, a police constable infected by *streptococci* and *staphylococci* bacteria after scratching his face on a rosebush. With a rampant infection of his face, head and lungs, PC Alexander was close to death.

Dr Fletcher administered one-fifth of a gram of Florey and Chain's penicillin to Alexander, then smaller doses every three hours. Within five days, Alexander had undergone an almost miraculous recovery and his wounds were healing. Tragically, they did not have enough penicillin to rid Alexander of infection. They even tried extracting it from his urine and giving it back to him, but it was not enough. The bacteria fought back, killing the policeman a few weeks later. However, the next five patients who were treated all made complete recoveries. The tests were an impressive demonstration of the powers of penicillin. The proof of principle had been made. The problem now was to mass-produce the substance so that more patients could benefit. In June 1941, Florey and Heatley flew to America to seek commercial assistance, but still there was no interest from industry or the American government. Six months later, everything changed as the attack on Pearl Harbor by the Empire of Japan brought the United States into the Second World War.

Realising that its troops would soon be going into battle, the American government poured resources into anything that showed promise of reducing medical casualties. With assistance from Florey and Heatley, two pharmaceutical companies developed methods of producing commercial quantities of penicillin by deep fermentation in large aerated vessels.

By 1942 there was enough of the drug to allow a completely conclusive trial on 187 cases, largely carried out by Florey's wife. The subsequent announcement of the successful tests brought a simmering resentment between Fleming and the Chain-Florey team bubbling to the surface.

Almroth Wright, Fleming's mentor and head of the inoculation department at St Mary's, wrote a letter published in *The Times* claiming priority for Fleming. The next day, Sir Robert Robinson, professor of chemistry at Oxford, responded that if Fleming deserved a 'laurel wreath' then Florey should be given a 'handsome bouquet'. Scenting a scrap among academics whose work was vital to the war effort, the press pitched in, but Fleming's humble life story, his long and lonely road to success, and the serendipity of the mould blowing in through his open window chimed better with the beleaguered wartime public than that of foreign-born state-funded researchers. From that moment on, Fleming assumed ownership of the discovery.

In the summer of 1943, Florey went to Tunisia and Sicily to find out how a small amount of penicillin could be used most efficiently for the treatment of war wounds, and by the time of the Normandy D-Day landings in 1944 there was enough of the 'wonder drug' to treat all the severe cases of bacterial infection that broke out among the troops on the drive to Berlin. By then, commercial production had begun in Britain, but only after Florey had been accused of giving away valuable commercial secrets to American companies that had subsequently patented technical aspects of the production methods. Until then, British researchers regarded the patenting of medical discoveries as unethical, a mindset that swiftly changed after the end of the war in 1945.

Hitler's war is said to have killed more than 60 million people; penicillin is believed to have saved more than 80 million. In the post-war years penicillin became much more than an antibacterial drug. An icon for post-war reconstruction, it brought an end to infections that had previously killed indiscriminately. Pneumonia, venereal diseases, diphtheria, scarlet fever and many other infections suddenly became treatable. With the inventors of nuclear power, the jet engine and the computer, Fleming was included in the roll call of scientists whose discoveries had shaped the twentieth century, although he referred to the fame he received for his contribution as the 'Fleming Myth' because his celebrity was the result of a fallacy that the discovery of penicillin was the fruit of a lone scientist's work. Fame in fact sat awkwardly with Fleming, who was so shy that talking with him was once described as like playing tennis with someone who, instead of returning service, always put the ball in his pocket. But this very reserve and humility now played to his advantage, increasing his popular appeal. Meanwhile, Florey and Chain were sidelined, their roles in the story overlooked by the general public, although all three men shared the 1945 Nobel Prize in Physiology or Medicine. Heatley was not included in that prize, despite the fact that, as Sir Henry Harris, erstwhile Regius Professor of Medicine at Oxford, later said: 'Without Fleming, no Chain; without Chain, no Florey; without Florey, no Heatley; without Heatley, no penicillin.'

Another overlooked vital contributor to the development of penicillin was Dorothy Hodgkin, a committed socialist and pioneer of X-ray crystallography, who, after being born in Cairo, strove throughout her professional and private life to improve education in Africa and bring together scientists to campaign for global political and social issues.

In 1945, Hodgkin solved the structure of penicillin by X-ray crystallography, thereby greatly aiding the development of new methods of synthesising the drug. Although her role in penicillin's development is not widely appreciated, she was awarded the Nobel Prize in Chemistry in 1964 (having been proposed at least twice previously) for her X-ray crystallography detection of several substances, including vitamin B_{12} and penicillin. Hodgkin was only the third woman to be so distinguished after Marie Curie and her daughter Irène Joliot-Curie, and she is the only British woman scientist to win a Nobel Prize to date. Four years later, she announced probably her greatest achievement, the determination of the structure of insulin – a discovery that led to her becoming the first woman since Florence Nightingale to be admitted to the Order of Merit.

* * *

Dame Hodgkin's solving of the structure of penicillin was notable also for her use of a Hollerith punched card calculator, the first time an electronic computer had been used to solve a biochemical problem. Made by the founder of the company that became IBM, Herman Hollerith, and based on punched card technology that was widely used throughout the nineteenth century to control textile looms and operate fairground organs, punched card calculators were one of the first devices used to process vast amounts of data. At the time, they were thought to be the most sophisticated and powerful calculating machines available, but that was only because the invention of a far more powerful machine was Britain's greatest secret of the Second World War.

That machine was the Colossus, developed at Bletchley Park near present-day Milton Keynes. A group of mathematicians, crossword compilers, chess players, statisticians, engineers and Egyptologists assembled by the government relied on Colossus and its predecessor, the Bombe, to unravel the codes used by the German military to send messages to field commanders and U-boats in the North Atlantic.

The Bombe's chief architect was Alan Turing, an eccentric and sometimes awkward man. Turing's stroke of genius was to wonder whether a machine could think and then to work out how it might emulate human thought. Thus, through his theoretical work in mathematics during the 1930s, Turing invented the concept of the computer as we know it now.

But Turing's best-known creation was his work with his code-breaking colleagues in the 1940s. Now regarded as the single most significant contribution to the Allied victory in Europe, for many years it was considered so secret that no one in the general public knew about it. In part this was because their work was so sensitive that some details are still classified. However, it was also because Turing was very much an outsider, gay at a time when homosexuality was regarded as an illness and homosexual acts as a criminal offence. Whatever the precise cause of his mysterious death in 1954, Turing's persecution because of his homosexuality shaped his fate in one of the most tragic stories to befall any genius of British science.

Like many original thinkers and scientific mavericks, Turing had an unconventional childhood and did not suit the constraints of orthodox schooling. Born into a highly traditional upper-middle-class family in 1912, Turing's parents lived in India until his father's retirement when Turing was in his teens. With his brother, John, Turing grew up in England, fostered by relatives and friends when he was not at boarding school. It has been suggested that Turing's loneliness and self-absorption as a child might have shaped his fascination with the operations of the human mind and the concepts of consciousness.

At preparatory school Turing already showed signs of mathematical genius (as well as an interest in maps, chess and debating), but mathematical skills did not conform with the classics- and humanities-centred curriculum of the between-the-wars public school system for which he was being prepared. Nevertheless, Turing developed a deep interest in science and nature, the turning point coming when he discovered *Natural Wonders Every Child Should Know*, a book that opened his eyes to the concept of scientific explanation. One chapter in particular, titled 'Where We Do Our Thinking' and concerned with the physiology and functioning of the brain, fascinated more than any others. When Turing wrote to his mother in his last year at prep school to say that he was 'making a collection of experiments in the order I want to do them in', she became extremely worried that he was neglecting his more traditional schooling, but her fears were assuaged when he passed his Common Entrance exams and was accepted for Sherborne School in Dorset.

Sherborne's start of term coincided with the General Strike of 1926. However, the 14-year-old Turing was so determined to attend his first day that he rode his bicycle unaccompanied more than sixty miles from Southampton to school, stopping overnight at an inn. Unfortunately his determination did little to impress the teachers after he arrived at the school. His poor handwriting, his 'dirty' and unconventional mathematical workings, his position at the bottom of the form in English and his messy experimentation in chemistry immediately attracted criticism.

However, these superficial shortcomings masked a fierce intellect. Within a year of arriving at Sherborne, Turing was solving advanced mathematical problems without ever having studied elementary calculus. The next year, aged 16, he read Albert Einstein's work on relativity; not only did he grasp it, but he extrapolated Einstein's questioning of Newton's laws of motion from a text in which it was never made explicit.

Despite his abundant and apparent abilities, Turing still failed to find favour with the school's authorities, his headmaster writing to his mother that 'if he is to stay at a public school, he must aim at becoming educated. If he is to be solely a scientific specialist, he is wasting his time at a public school.' With such attitudes commonplace at the time, it is a small miracle that empire-era Britain produced any notable scientists at all. In spite of his headmaster's displeasure and concern that he would not pass his exams, Turing continued his private study of the key works of relativity, quantum mechanics and mathematics. His private notes on the theory of relativity showed a self-taught degree-level appreciation, yet he was almost prevented from taking the school certificate lest he shame the school. Then, in the sixth form, school provided Turing with an experience that shaped his life and all of mankind's. Turing broke out of his solitary shell when he fell in unrequited love with a slightly older boy, Christopher Morcom, who shared his interest in science. He dreamed that they might go up to Cambridge together. Christopher duly won a scholarship in 1929. Then tragedy struck. Only a few weeks after the two boys entered their last term at Sherborne, Morcom died suddenly from complications arising from bovine tuberculosis contracted after drinking infected cow's milk as a young boy.

Turing's religious faith was shattered by the death of his friend and he became an atheist, adopting the conviction that all phenomena, including the workings of the human brain, must be materialistic. With no soul in the machine and no mind behind the brain, how then, Turing wondered, did thought and consciousness arise? In his anguish, he wondered whether there was any way in which Christopher's mind might have survived the death of his body. What was the relationship between a mind and a body – between mind and matter? Answering those questions would shape his years at university and drive his determination to understand the nature of thought.

Turing could not resurrect his dead friend, but there was something he could do: win a scholarship to read mathematics at Cambridge, just as Christopher Morcom had. Having neglected his classical studies, Turing narrowly missed winning a scholarship to Trinity College, Cambridge, a lodestar at the time for mathematicians. Instead he went to King's College, Cambridge, where he thrived under a free and tolerant environment, acquiring his lifelong interest in athletics and graduating with a first class honours degree in 1934. Over a period of five years, he produced a series of papers that investigated probability, logic and several other arcane, deeply technical aspects of the philosophy and limits of mathematics.

King's College suited Turing. It tolerated homosexuality at a time when homosexuality in Britain was not just frowned on in public, but actually illegal.

At King's, Turing was at last taken seriously academically. He was elected to a fellowship at the age of only 22. But he wasn't just a scholar; he also liked rowing and running.

One day in 1935, out running through Grantchester Meadows, Turing had one of the greatest mathematical insights of all time. He had been thinking about the theoretical limits of mathematics. His great insight was to see how an imaginary machine operating with very simple rules but an infinite amount of time could solve any conceivable mathematical problem.

Richard Dawkins

Turing outlined his ideas about a machine that could solve any conceivable mathematics problem in his greatest work, *On Computable Numbers, with an Application to the Entscheidungs-problem*, published in 1936. Investigating whether there could exist, at least in principle, any definite method by which all mathematical questions could be decided, Turing said the key question was a definition of method. In answering this question, Turing outlined the theoretical concept of a computer: a machine that could perform any task provided that the task in question could be set out in a methodical sequence of mathematical algorithms. In other words: not a real machine, but what we now think of as a program. Turing called this machine a Universal Machine, although it is nowadays more usually referred to as a Turing Machine.

The Turing Machine was an entirely theoretical construct – a sequence of instructions that responded to inputs. But it introduced the idea that interpreting the instructions and carrying them out was a mechanical process. Therefore any Turing machine could be made to do whatever any other Turing machine could do simply by supplying it with the same instruction sequence. One machine, for all possible tasks: the modern computer program. And the mechanical task of interpreting and obeying the program was what the modern computer itself now does.

THE EXPEDIENCY OF WAR

> All computers – whether real or imagined, whether now or in the future – must obey the mathematics set out in Turing's now-famous paper.
>
> The full power of Turing's revolutionary ideas would not be appreciated for years, even decades, but Turing was soon involved in something that would take him away from the rarefied beauty of pure mathematics and closer to the theoretical machine he envisaged.
>
> Richard Dawkins

All Turing now needed was a mechanical device capable of performing his instruction sequences. In principle that machine had been devised more than a century earlier by a 21-year-old Cambridge student called Charles Babbage.

If Alan Turing was the father of the computer, then Charles Babbage deserves the title of grandfather. Born in London in 1791, Babbage was the son of a wealthy banker, inheriting in his mid-30s a fortune which, in today's money, would be worth more than £5 million. It was always clear, however, that Babbage would not be following his father's career. As a maths student at Cambridge, his non-conformist attitudes led him to become a founder of the Analytical Society, dedicated to introducing modern continental mathematics into the 'moribund' university syllabus. Later he would turn his attentions on the Royal Society, supposedly the world's leading scientific body, by launching an attack on its indolence.

Born ahead of his time – Babbage once said he had never spent a happy day and that he would gladly give up the rest of his life if he could spend just three days living 500 years in the future – he arrived at Cambridge in 1810 and immediately discovered he knew more algebra than his tutors.

At that time, all calculations beyond simple addition and subtraction were performed using the logarithm tables invented by John Napier in the sixteenth century. In addition, seafarers relied on astronomical tables, surveyors needed trigonometry tables, fishermen used tide tables and bankers used compound interest tables. But all these tables were riddled with inaccuracies. The preparation of such tables had become a cottage industry. Rural clergymen, who were well educated and had time to spare, oversaw their calculation by armies of 'computers'. These human adding machines, often teenagers, each worked on a small part of a formula for the quantity in question. Their individual efforts would then be pooled to give the final figure for entry in the mathematical tables.

It was a recipe for errors, and they duly appeared in their thousands. By 1784, fears about the reliability of mathematical tables reached a level where the French government decided to start from scratch, commissioning the Baron Riche de Prony to prepare a definitive set, calculated to 10 decimal places. Compiled by a team of around 100 mathematicians and computers, many of them unemployed hairdressers who, before the revolution, had tended the elaborate coiffures of the aristocracy, the results filled seventeen handwritten volumes packed on every page with tables. Yet they too were shot through with errors.

When Babbage was idly contemplating such tables one evening in 1812, he realised that the errors were generated by the hairdresser-computers. Only needing to know how to add and subtract, their calculations were based on a straightforward but laborious calculation process called the method of differences, so Babbage decided to replace them by a machine – the Difference Engine – which used the same method, albeit without human mental effort.

By performing the calculations mechanically on a series of gears and axles, the Difference Engine would not make mistakes. To eliminate errors of transcribing the results of the calculation, Babbage's Difference Engine would print out the results.

It took ten years, but by 1822 Babbage had a small prototype working. It was clear, however, that the full-sized Difference Engine, with its estimated 25,000 parts, would require far more effort and expenditure. After winning government backing, Babbage set to work again, but almost another decade later he had only a fraction of the Difference Engine completed. Despite working perfectly, it failed to impress the government, as did Babbage's revelation that he was working on plans for an entirely new device that would make the Difference Engine obsolete.

Known as the Analytical Engine, this locomotive-sized machine would act like a mechanical mathematician, actually solving equations. Astonishingly, it featured components now found in every personal computer, such as a mechanical memory and a central processor. It could be programmed using punched cards, an idea Babbage borrowed from the textile industry. His friend, the amateur mathematician Ada, Countess of Lovelace, daughter of the poet Lord Byron, even devised a program for calculating so-called Bernoulli numbers using the device, and thereby became the first computer programmer.

Yet the staggering complexity of the Analytical Engine (plus Babbage's constant revisions of its design, and the understandable scepticism of the government) ensured it was never built. Nothing approaching its power or versatility would be seen for over a century.

Even so, Babbage's vision of performing specialised calculations using machines came to fruition within two years of his death. In 1873, William Thomson (a.k.a. Lord Kelvin, who devised the laws of thermodynamics) unveiled an ingenious arrangement of pulleys, gears and axles, the movements of which sketched the rise and fall of tides on a rotating drum. Over the next decade, a variety of these tide prediction machines were devised, their accuracy only finally being superseded by electronic computers in the mid-1960s.

The next step towards Turing's concept of a Universal Machine took place in the 1880s. Faced with a vast wealth of data produced by the US census, an American statistician named Herman Hollerith hit upon the crucial idea of using a machine for data analysis. In 1884, he filed the first of many patents for electro-mechanical devices that turned data into punched cards, and then sorted them into categories. Used to analyse the results of the 1890 US census, Hollerith's machines produced a basic analysis in just six weeks, three times faster than before. It was an ancestor of one of these machines that helped Dorothy Hodgkin determine the structure of penicillin.

Neither the punched card calculators used by Hodgkin nor the machines that Turing helped to develop during the war were strictly speaking Turing Machines. They were designed to perform specific mathematical tasks and could not be programmed to undertake any chosen assignment: the true definition of a Universal Machine. However, they were the final steps along the long road to the modern computer, and the last machine that Turing helped to develop was in effect the first digital computer.

By the late 1930s, British cryptographers were starting to realise that Turing's idea of a Universal Machine might be ideally suited to code breaking. Recognising Turing's talents and usefulness to the government, military intelligence decided to ignore the potential security risk of his sexuality. While still at Cambridge, Turing secretly started working part time for the Government Code and Cypher School, the British code-breaking organisation.

Until Turing's arrival in 1938, the GCCS had not a single mathematician or scientist among its members, the authorities believing that code breaking was the province of arts specialists. For years, very little progress had been made in deciphering German military communications. However, that changed with the arrival of Turing and the passing on of information and advice from Poland, where mathematicians had achieved great successes in cracking German codes.

The day after Britain declared war on Germany, Turing reported at Bletchley Park, a Victorian mansion that housed the wartime cryptanalytic headquarters, codenamed Ultra but known to insiders as Room 47 of the Foreign Office or Station X. Five weeks before Germany invaded Poland, Polish intelligence had given their British and French counterparts reconstructions of the Enigma machine used by the German military to encode its communications. They also passed on techniques for decrypting ciphers produced on Enigma and details of the 'Bomba' device that they used to mechanise and speed up the code-breaking process.

Masquerading as Captain Ridley's Shooting Party to disguise their true identity, Britain's best code-breakers set to work at Bletchley Park. Turing immediately set about developing a more powerful version of the Bomba to break German codes quicker. Although the Allies knew how the Enigma machine worked, they needed to know how its three rotors (four in the case of machines used by the German navy from 1942) had been set before they could crack the codes. The starting position of the rotors, each of which could be set to one of 26 positions, was changed at midnight every day and then for each message. By adding other variables, such as a plug board, an Enigma machine could be set to one of around 100 billion billion permutations, a number that makes winning the lottery, for which the odds are a mere 14 million to one, appear by comparison almost a certainty.

To crack the ciphers produced by the formidable Enigma, Turing developed the Bombe, a contraption described by its operators as looking 'like great big metal bookcases'. It was about seven feet (2.1m) wide, six feet six inches (2.0m) tall and two feet (0.61m) deep. With space for 108 six-inch rapidly rotating drums on its front, it weighed about a ton.

Provided the operators had a 'crib', a snippet of encrypted text that they had managed to decode, the Bombe could then use brute force to work out each possible rotor position on the Enigma machine and encrypted messages could then be broken. The operation was based on Turing's idea that contradictions could tell him everything: by finding rotor permutations that did not work with a crib they could eliminate that setting and often several others until the correct permutation emerged.

To find a crib, the code-breakers looked for frequently used phrases, such as 'nothing to report' or a standard greeting. When stumped, Bletchley Park would sometimes ask the Royal Air Force to seed an area in the North Sea with mines (a process that came to be known as gardening, for obvious reasons) and then wait in hope that the Enigma messages sent out shortly afterwards would contain the name of the area threatened by the mines, which could be exploited for the crib.

Turing's first two Bombes were in operation in 1940 and decoded 178 messages. By the end of 1941 sixteen Bombes had been constructed and tens of thousands of coded messages had been cracked. Thereafter, the number of Bombes in use more than doubled each year until by May 1945 there were 211 operational machines, requiring nearly 2,000 staff to operate them. With their drums spinning at 120 revolutions per minute, the Bombes took an average of around three hours to break a typical Enigma message, providing the Allies with access to a large part of German military communications.

The prize for breaking the Enigma codes was survival. The U-boat Wolfpacks in the North Atlantic were sinking convoys every day and severing the British lifeline to America.

All U-boat secret communications used Enigma, but with Enigma cracked the Atlantic convoys could be diverted to avoid the Wolfpacks.

What the U-boat commanders had called their 'happy time' had reached its end.

The battle of the North Atlantic could not have been won without Alan Turing and his code-breakers at Bletchley Park.

Richard Dawkins

However, in February 1942, the German navy changed the Enigma machines used by the Atlantic U-boat squadrons, so Turing called in telephone engineers to exploit their electronics skills to make the Bombes work faster.

By this time, Turing was cutting an eccentric figure around Bletchley Park. Shabby, nail-bitten and tieless, he lived in a pub outside Bletchley and would cycle to work wearing a gas mask to combat his hay fever. Once ensconced in hut 8, where German naval codes were tackled, he chained his tea mug to a radiator for fear that someone would steal it and set about promoting an atmosphere of creative anarchy. Although sometimes halting in speech and awkward of manner, he was the source of many hilarious anecdotes about bicycles, gas masks and the Home Guard, and was regarded as the foe of charlatans and status-seekers. Relentless in long shift work, he would often arrive at midday and then work until midnight the following day. When the problem was solved, he would disappear and rest for twenty-four hours. Seeking relief from the stresses of Bletchley Park, he would undertake long-distance runs, often running more than twenty-five miles to meetings in London.

In 1943, Turing aided the development of Colossus, the world's first programmable digital electronic computer. It was used to help decipher German higher command teleprinter messages, which had been encrypted using a new Lorenz SZ40/42 cipher machine. It generated more than a million times more combinations than Enigma. The Bletchley team built ten Colossus machines especially to decipher Lorenz signal traffic between Hitler and his generals before the Allied invasion of Europe. It was so successful that by D-Day the code-breakers could read the German orders of battle and assure Eisenhower and Churchill that Hitler had been fooled by their plans. Hitler was expecting the invasion on the Channel coast, not the Normandy beaches. Between them, the Colossus and Bombe computers are credited with providing the Allies with sufficient information to shorten the Second World War by at least two years.

After the war, Turing returned to Cambridge, hoping to resume his quiet academic life, but the newly created mathematics division of the British National Physical Laboratory offered him the opportunity to create an actual Turing machine, the ACE or Automatic Computing Engine. Turing accepted, but found that red tape and interminable delays had replaced the wartime spirit that had short-circuited so many problems at Bletchley Park. Finding most of his suggestions dismissed, ignored or overruled, Turing returned briefly to Cambridge before accepting an offer from the University of Manchester, which was also proposing to build a proper Turing Machine, the Manchester Mark 1.

The end of the war also gave Turing time to focus on his cross-country running. He would amaze his colleagues by running to scientific meetings, beating the travellers by public transport. A frequent top-rank competitor in amateur athletics meetings, only an injury prevented his serious consideration for the British team in the 1948 Olympic Games.

Meanwhile in the lab, Turing had broadened his thoughts on thinking machines since his publications in the mid-1930s. He now proposed a machine that could learn from and thus modify its own instructions, the first suggestion of what came to be called artificial intelligence. In an article published in 1950 in *Mind*, a philosophical journal, Turing proposed what he called an 'imitation test', later called the Turing test. It involved an interrogator in a closed room asking questions of two subjects, one human and the other a computer, both of which he could not see. The interrogator had to decide which responses came from the human and which from the computer. If he or she could not do so, then the computer could be said to be 'thinking' as well as a human. Turing remains a hero to proponents of artificial intelligence, in part because of his assumption of a rosy future: 'One day ladies will take their computers for walks in the park and tell each other, "My little computer said such a funny thing this morning!"'

Unfortunately for Turing, reality caught up with him well before his vision was realised. In Manchester, he told police investigating a robbery at his house that the burglar could very possibly be an acquaintance of a man with whom he had recently started 'an affair'. Always frank about his sexual orientation, Turing this time got himself into real trouble.

Turing was tried and convicted of gross indecency in 1952, the same charge and conviction that had famously sent Oscar Wilde to jail nearly sixty years earlier. Turing was spared prison but subjected instead to chemical castration involving injections of female oestrogen hormones intended to dampen his lust. 'I'm growing breasts,' he told a friend.

Although Turing continued to work part time for GCHQ, the post-war successor to Bletchley Park, the conditions of the Cold War and the alliance with the United States meant that convicted homosexuals had become ineligible for high-level security clearance. Having already frazzled the nerves of the security services by taking a holiday in Greece in 1953, he added to the acute anxiety at that time about spies and homosexual entrapment by Soviet agents due to the recent exposure of the first two members of the Cambridge Five, Guy Burgess and Donald Maclean, as KGB double agents.

On 8 June 1954, Turing's cleaner found him dead. Lying beside him was a half-eaten apple said to be laced with cyanide. A post-mortem established the cause of death as self-administered cyanide poisoning, although conspiracy theorists continue to suggest that he was assassinated because of the security risk posed by his homosexuality. It has subsequently been reported that the apple was never tested for cyanide contamination. His mother always maintained that Turing's death was an accident incurred while conducting electrolysis experiments. He was 41.

* * *

The career of Frank Whittle, whose invention of the jet engine was, like the Turing machine, initially ignored before being seized upon during the war years, was altogether more successful, although not without its disappointments.

Whittle ended up an Air Commodore in the Royal Air Force and lived to the age of 89, by which time he had seen his creation become one of the great life-transforming inventions of all time, bringing long-distance travel to the masses. However, his success was tinged with some bitterness. He had to overcome years of obstruction from the authorities, and a lack of funding and faith in his brilliant ideas. With justification, Whittle felt that if he had been taken seriously earlier, Britain would have been able to develop jets before the Second World War started. Had that happened, it is reasonable to assume the war would have been considerably shortened and millions of deaths potentially avoided.

> If Whittle had been listened to several years earlier, then Britain could have had jet aircraft flying in the Battle of Britain in 1940.
>
> What fascinates me is why nobody believed the eminently believable Frank Whittle, a true genius and one of the greatest inventors ever, not that anyone in officialdom at the time was bright enough to notice.
>
> Some say that the problem lay in Whittle's working-class background. Others say his single-mindedness was at fault, but whatever the cause Whittle thrust Britain into the jet age and turned the aviation industry on its head – all with zero encouragement and quite a bit of opposition from the government.
>
> James Dyson

Born in 1907 to a skilful and inventive foreman in a Coventry machine tool factory, Whittle's fascination with aircraft and flying began at the age of four when his father gave him a toy aeroplane with a clockwork propeller. By the time Whittle was nine, his family had moved to Leamington Spa, where his father, Moses, had bought the Leamington Valve and Piston Ring Company and the young Whittle first learned his mechanical skills. 'It was while helping him that I first acquired practical experience of certain manufacturing processes, when I was only ten years old,' he later said.

The sight of aircraft overhead (including on one occasion force-landing near his home) and being built at a local works during the First World War fed his obsession with flying, which he supplemented by poring over popular science and engineering books in his local library.

As soon as he was old enough, in 1923 Whittle applied for an apprenticeship at RAF Halton. 'I passed the written examination with flying colours,' he said, 'but failed the medical examination because, being only five feet tall, I was considered to be undersized.'

Refusing to let his stature, fitness and small chest measurement deter him, Whittle embarked on a gruelling training programme that saw him gain three inches in height and add a further three to his chest measurement in only six months. But still it was not enough. Rejected again, he was barred under RAF recruitment rules from making a third attempt, so he reapplied under a different name, passed the written test again and was accepted for an apprenticeship at RAF Cranwell, 'having got in under false pretences in a sense'.

Despite not enjoying his apprenticeship, Whittle passed out sixth of 600 apprentices, but only the top five qualified for coveted flying cadetships at the RAF College. Fortunately for Whittle, the top apprentice failed his medical. Whittle's commanding officer strongly recommended him and after just eight hours' instruction he was showing his innate talent by flying solo in an Avro 504N biplane.

Whittle graduated to Bristol fighters and developed a reputation as something of a reckless daredevil. On one occasion, having got lost due to poor visibility and having no compass fitted to his aircraft, he landed in a field to ask directions, then decided to attempt a takeoff and struck a tree, writing off the aircraft. Later on, disqualified from an aeronautics prize for having performed a disapproved 'bunt' manoeuvre, he was also almost court-martialled after a public complaint about his low flying and aerobatics.

Before graduating from Cranwell, Whittle wrote a revolutionary thesis, *Future Developments in Aircraft Design*, in which he suggested that the next generation of aircraft could achieve speeds of 500 m.p.h. by flying higher and using rocket propulsion and gas-turbine-driven propellers high up in the stratosphere, where the air was too thin for conventional piston-driven propeller engines. Although he did not mention the use of a gas turbine for jet propulsion, he continued to consider the problem and in 1928, during his spare time while participating in an RAF flying instructor course, he came up with the idea of using a gas turbine to produce a jet-powered aircraft. The principle of using a jet of hot air to thrust an engine forwards was beautiful and simple. It obeyed Newton's third law of motion that every action produces an equal and opposite reaction.

Whittle's ideas must have seemed like science fiction or fantasy, but he had done the mathematical calculations to prove them. He was awarded full marks for his thesis, with his professor remarking, 'I couldn't quite follow everything you've written, Whittle, but I couldn't find anything wrong with it.'

James Dyson

Whittle's design had a gas turbine driving a compressor to suck vast quantities of air into the engine. Within the engine, fuel would be injected into the air and ignited, causing it to expand extremely rapidly and escape through the engine's jet exhaust. This vast mass of hot, fast air escaping out of the back of the engine would, according to Newton, cause the equal and opposite reaction of pushing the engine (and the aircraft to which it was attached) forwards at high speed. And the faster the jet moved through the sky, the more air it would suck in and therefore the greater would be its potential forward thrust. To Whittle's great fortune, one of his instructors, Flying Officer Pat Johnson, had been a patent agent in civilian life. He recognised the potential of Whittle's idea and immediately arranged an interview with the commandant.

Within days, Whittle was on his way to meet Dr A.A. Griffith of the Air Ministry's South Kensington laboratory, but Griffith dismissed Whittle's design, insisting that gas turbines were practicable only for driving propellers. Whittle returned to his instructor course and qualified, meanwhile patenting his jet turbine design in 1930 with Johnson's assistance.

Whittle was only 22 years old when he filed his patent for a jet engine in 1930, but his design was so simple and so elegant that even today it takes my breath away.

His great idea was to design an engine with only one moving part instead of the hundreds of moving parts of conventional piston engines.

At the time, the legendary Rolls-Royce Merlin engine that went on to power the Spitfires and Hurricanes in the Battle of Britain was the best engine we had. It was magnificent, but it was inefficient and at the limit of what pistons and propeller could do.

Whittle's propeller-less jet turbine would drive a plane through the air by thrust alone. So simple!

James Dyson

The next year Whittle was posted to the Marine Aircraft Experimental Establishment at Felixstowe, where as a test pilot of floatplanes he encountered the most hair-raising escapade of his flying career.

Tasked with testing a seaplane catapult on HMS *Ark Royal*, Whittle found on takeoff that his open-topped plane veered almost out of control. When he looked over his left shoulder to check whether his aircraft had been damaged by the launch, he saw a person lying face down on top of the plane. Checking over his shoulder again, he realised it was his passenger, Flight Lieutenant Fred Kirk.

With great presence of mind, Whittle landed the barely controllable seaplane, narrowly avoiding a German liner in the process, and Kirk clambered into the rear cockpit. 'He seemed quite calm and collected, while I was almost in a state of nervous collapse,' Whittle said afterwards.

Shaking with nerves, Whittle exclaimed to Kirk: 'My God!'

'What are you worrying about?' replied Kirk.

Pink gins were pressed into Whittle's and Kirk's hands by several very relieved naval officers immediately after they were hoisted aboard HMS *Ark Royal*. Kirk then explained what had happened. At takeoff Kirk had lost his grip on the rim of the cockpit because his gloves were slippery with glycerine. Bounced into the air and spinning around, Kirk had kept his cool sufficiently to grip the fin wire before any other part of him touched the plane. He then held on for his life.

Shortly afterwards Whittle was sent on an officer's engineering course at RAF Henlow and performed so well that he was allowed to enrol on the Mechanical Sciences Tripos at Cambridge. He graduated with a First in 1936 and intended to pursue postgraduate research into his jet engine, although by then his patent had lapsed after the Air Ministry refused to pay the £5 renewal fee.

Whittle was about to give up on his jet hopes when Ralph Dudley-Williams, a friend from his Cranwell days, and James Collingwood Tinling, a former RAF pilot, approached him, offering to act as his agent and to cover the expenses of further patents as well as research and development costs. With City backing, they formed a company called Power Jets, and the Air Ministry gave

Whittle permission to act as honorary chief engineer provided it did not conflict with his official duties.

Using new alloys to construct the combustion chambers, Whittle had a test engine ready by April 1937. It looked like a cross between an old gramophone trumpet speaker and a vacuum cleaner, but it worked, albeit with frightening levels of noise and vibration. On 12 April it was tested successfully, although it soon raced out of control, causing the test team to bolt for cover. With financial support from the Air Ministry, Power Jets was given a new test site at Lutterworth, near Rugby, and the engine was rebuilt with ten combustion chambers instead of the previous single chamber.

At last the RAF top brass started to take a close interest, although not always without incident. When Air Chief Marshal Sir Hugh Dowding, Commander-in-Chief of Fighter Command during the Battle of Britain, visited the works, Whittle demonstrated one of his engines in operation.

'I pointed to the nozzle,' Whittle later said, 'meaning to imply: "That's the business end of the engine."'

'He misunderstood my gesture and, before I could stop him, walked rapidly in the direction indicated. Suddenly a mighty invisible force wrenched open his raincoat and sent him staggering across the concrete, his brass hat rolling away on the grass.

'I stood petrified with horror, and when Sir Hugh recovered himself, apparently unhurt, I could scarcely move and certainly could not speak.'

Whittle's jet was immediately switched off.

'Sir Hugh, pardonably terse, asked me if I wasn't going to show him something, so I dumbly led the way into the test house. Only those in the Service would really understand what a junior Wing Commander feels like when he is responsible for this sort of thing happening to an Air Chief Marshal.'

In spite of the Air Chief Marshal's pratfall, the Air Ministry now realised the potential of Whittle's engine. And not a moment too soon. In Germany, the 22-year-old physicist Hans Joachim Pabst Von Ohain had patented a jet engine in 1933. Unaware of Whittle's work, Von Ohain had designed his engine slightly differently but the concept was the same. With support from the Nazi government, a prototype was built and tested. In August 1939 a small aircraft, the Heinkel He178, was fitted with Von Ohain's engine and the first flight of a jet-powered aircraft took place, beating Whittle into the air. However, the plane was a dead end. It could fly for only six minutes and the engine had to be completely rebuilt after each flight, so after some promise the He178 was scrapped, much to the relief of the Air Ministry. By contrast, Whittle's engine was a triumph from the start.

With the outbreak of war in September 1939, the Air Ministry commissioned a more powerful engine from Whittle, which it wanted fitted to an experimental aeroplane, the E28/39, built by the Gloster Aircraft Company. Nevertheless, a series of bureaucratic betrayals followed, such as the Air Ministry demoting Whittle's Power Jets to a research organisation and then offering shared production and development contracts to rival companies, which were allowed to make alterations to his design behind Whittle's back. Drinking and smoking heavily to cope with the stress, Whittle was losing control of his creation, but t 7.40 p.m. on 15 May 1941 he watched as the *Gloster* – a plane without a propeller – made its maiden flight at RAF Cranwell. Flying low and fast at 340 m.p.h., it was the test flight of Britain's first jet, but the Air Ministry was so uninterested it hadn't even bothered to send a photographer. Luckily an amateur cameraman captured the historic event on film.

As it flew overhead on the first of ten hours' successful flight trials, Johnson patted Whittle on the back. 'Frank,' he said, 'it flies.'

'Well, that was what it was bloody well designed to do, wasn't it?' replied Whittle.

He later said that by the time of the first test flights he had been 'waiting for this for so long that I can't honestly say that there was a very great thrill attached to it. I just knew that it would fly. No reason why it shouldn't.'

Whittle's intellectual secrets were shared with several British and American companies, including De Havilland, Metropolitan-Vickers and General Electric. Rolls-Royce took over production of the next generation of engine from Power Jets.

Fearful that his creation would be eviscerated by private aircraft companies, Whittle suggested that the programme to build a jet fighter should be nationalised. Sir Stafford Cripps, the Minister of Aircraft Production, immediately took Whittle up on it, turning

Power Jets into a state research organisation and leaving Whittle with nothing financially, although he was promoted to air commodore and had the satisfaction of knowing that Gloster Meteors powered by his jet engines were shooting down German V1 flying bombs.

The Meteor was beaten into the air as the first operational jet by a German fighter-bomber, the Messerschmitt Me-262, but the end of the war was in sight and the Me-262 had negligible impact on its outcome. In 1948, Whittle retired from the RAF having been knighted, given a £100,000 award and elected to the fellowship of the Royal Society. He moved to America to take up a university appointment, from where he watched his engines quite literally make the world a smaller place after they were employed on a succession of passenger jets, starting with the ill-fated De Havilland Comet and culminating with the glorious Aérospatiale-BAC Concorde.

Many years after he developed the jet engine, Whittle became good friends with his great wartime rival, Hans Von Ohain.

Von Ohain then remarked that if British officials had backed Whittle, World War Two would never have happened. Hitler would have doubted the Luftwaffe's ability to win.

If only those short-sighted and dim-witted Ministry bureaucrats had recognised the brilliance of Whittle's amazing invention.

James Dyson

Chapter 14

THE MEANING OF LIFE

When Britain emerged from the Second World War, it looked like British science, which had achieved extraordinary things in the first half of the twentieth century – from splitting the atom and devising the concept of the computer to antibiotics and the jet engine – might be about to lose its pre-eminence in the world. With resources extremely limited and the economy recovering only slowly from the devastation of war, there was every reason to assume that science would be a low priority for funding and support.

But instead of slowing down, something rather amazing happened to British science. It took off again. The story of the last fifty years has been one of astonishing breakthroughs in which some of the really big questions have been answered. Questions such as how did the universe begin? Where do we come from? And how are we made?

Without doubt, the most significant event in post-war science and the most important biological advance of the twentieth century was the discovery of the structure of DNA, the molecule that exists in every living cell and which makes each of us so unique. It was the discovery of the key to what we are.

The unlocking of the structure of deoxyribonucleic acid, to give the molecule its full name, was one of the most thrilling break-throughs in the history of science. And like all good thrillers, it was a story of rival teams racing towards a goal, with the winner not necessarily the most deserving competitor, but the most ruthless.

Most people associate James Watson and Francis Crick with the discovery of the structure of DNA. Some know that they were not the only team to make the discovery, but few realise that their discovery was not the result solely of their own work. In fact, Watson and Crick achieved their breakthrough only after gaining a sneaky peek at their rival's findings. But chance favours the brave in science as much as in any other walk of life, so the garrulous and audacious Watson and Crick captured the plaudits. Helped considerably by Watson's account of the race in *The Double Helix* (a book that was so one-sided and dismissive of rivals that Watson's publisher postponed its publication until he had toned down some of its more derogatory comments), the public remember Watson's and Crick's names before any others.

> They were an unlikely pair, not what Cambridge was used to. Watson and Crick were irreverent, arrogant even, searching for nothing less than the secret of life, what Watson himself referred to as 'the most important event in biology since Darwin'.
>
> Richard Dawkins

Watson was a flashy and intuitive young American, a child prodigy chosen to appear on US radio as a 'quiz kid' to demonstrate his high IQ and prodigious knowledge. With a fascination for biology inspired by going bird-watching with his father, Watson had entered university at 15 and graduated at 19, arriving at Cambridge at 23, bursting to make his name.

At 35 Crick was much older, but just as clever and larger than life. Like Watson, he had also been inquisitive and precocious as a child – so precocious that he worried that by the time he grew up there would be nothing left to discover. Watson found him inspiring when they met in Cambridge in the autumn of 1951 and they readily agreed to team up to investigate DNA, becoming a team unlike anything Cambridge's science faculty was used to: loud, arrogant and wildly experimental.

Neither Watson nor Crick had a formal background in the newly emerging subject of biochemistry. Having switched from particle physics, Crick was still working on his doctorate in organic chemistry and X-ray crystallography, while Watson hoped, he said, 'that the gene might be solved without my learning any chemistry'. Instead they intended to use logic, lateral thinking, model building – and cunning – to deduce the structure.

If they could thereby discover the secrets of inheritance, it would be, as Watson put it, 'the most famous event in biology since Darwin's book'. And they were determined to get there before anyone else.

By the time that Watson and Crick started their work in 1951, a lot was already known about DNA and its position at the heart of every cell. In the seventeenth century, as we noted earlier, Robert Hooke had coined the term 'cell' and suggested that these might be the building blocks of all living organisms. Then in 1831 the Scottish botanist and protégé of Joseph Banks, Robert Brown (after whom Brownian motion is named), had discovered that every plant cell has a central structure, which he called the nucleus from the Latin for 'little nut'.

By the latter half of the nineteenth century, the capability of microscopes had improved sufficiently to allow scientists to observe the fundamental moment of reproduction taking place when a sperm penetrates an egg cell. From this it was known that the nuclei of the sperm and egg cells fuse together to create a new single-celled organism that then grows by dividing itself repeatedly.

Over the next few years of the 1880s, various scientists in Germany and Belgium watched as thread-like structures in the nucleus divided and duplicated themselves before each cell split. They also noticed that, in the case of sexual reproduction, threads from two parents come together in a single nucleus. Although unproved, the implication was clear: these threads, which they called chromosomes, might play a role in passing on hereditary features.

Meanwhile, in the 1850s and 1860s a Bohemian monk called Gregor Mendel performed a series of experiments on pea plants that he grew in his monastery's garden. From these experiments, he worked out the basic laws of heredity for all species that repro-duce sexually. He discovered that each parent passed on a factor related to a particular trait, for instance eye colour, which combined with a factor for the same trait from the other parent. Exactly how the trait would be seen in the next generation (or whether it might remain hidden until a later generation) depended on the combination of factors from each parent, and whether each factor was dominant or recessive. With each parent passing on a gene to its offspring, the only time that a recessive gene, such as blue eyes, would be expressed in the offspring would be when both parents passed it on. In any other circumstance – a dominant gene (such as brown eyes) from one parent and a recessive gene (such as blue eyes) from the other, or dominant (brown eye) genes from both parents – the dominant gene will always win over. This discovery fitted in exactly with Darwin's theory of evolution by natural selection, which it predated, but it was published in an obscure Austrian horticultural journal. Sadly, Darwin, who thought a baby inherited a blended mixture of vital fluid from its parents, remained unaware of how Mendel's research added evidence to support his theory.

The culmination of all this research was that by the turn of the twentieth century scientists knew the means and principle by which each of us becomes who and what we are. All that remained to be discovered was the mechanism.

Some scientists – but certainly not all – believed that DNA played some kind of role in the mechanism of heredity. It had been discovered that there were masses of it in every nucleus, nearly 2 metres in the case of humans, and that chromosomes were made of DNA, so it seemed a likely candidate.

DNA had first been discovered in 1869, in some pus-soaked surgical bandages examined under the microscope by a Swiss-German scientist, Johann Friedrich Miescher. Over the following years scientists worked out that it was made up of nucleic acid, phosphate groups, ribose sugars and molecular fragments called bases. These bases came in four varieties – adenine, cytosine, guanine and thymine – and they varied in number between DNA molecules, suggesting that their relative quantities might some-how affect the passing on of traits. The core fact that chromosomes are responsible for passing on genetic traits was confirmed by experiments on fruit flies, which with only four chromosomes are easier to study than most other animals. Then in 1944 the Canadian bacteriologist, Oswald Avery, proved that DNA was not a passive molecule but very much the *active* genetic agent, a discovery that was lauded as 'Avery's bombshell'.

But still no one could work out how DNA's four inauspicious bases might pass on complex genetic information. A further clue came in 1950, just a year before Watson and Crick started their work. Erwin Chargraff – an Austrian Jewish biochemist who had moved to the United States to escape Nazi persecution – determined that, although the amounts of the four bases might vary between two molecules of DNA, they obeyed two rules: the amount of cytosine was always the same as the amount of guanine; and similarly, the amount of adenine matched the amount of thymine. This suggested that the bases operated as C-G and A-T pairs. Nevertheless, the question still remained: how does an inanimate molecule replicate itself in such a way that it passes on the characteristics of life?

The answer, it was suspected, was hidden somehow in the architectural structure of DNA. If that conundrum could be cracked, maybe everything else would fall into place. The race was on.

In Britain, only two teams were deemed to have the skills and equipment to do the job, one at King's College in London headed by Maurice Wilkins and the other at the Cavendish Laboratory in Cambridge (not necessarily Watson and Crick, incidentally). To prevent a duplication of efforts and a waste of precious resources, the Medical Research Council insisted that Wilkins' team got the first crack. In no uncertain terms, Crick was told to concentrate on his PhD.

Wilkins was already a highly regarded scientist. Born in New Zealand in 1916, he had come to England at the age of six. After studying physics at Cambridge, he had risen through the research ranks steadily until the Second World War, during which he was a key member of the teams involved in developing radar and the British contribution to the Manhattan Project, the American-led programme to build the first atomic bomb.

By the end of the war, Wilkins had already achieved enough to cement his reputation. But disillusioned by the contribution of physics to the horrors of Hiroshima and Nagasaki, he shifted his interests in 1945 to the fast-growing field of biophysics and almost immediately started work on DNA: a science of life, not death.

The new biophysics department at King's was a ramshackle affair. Wilkins' team had to beg, borrow and steal their equipment. Their X-ray cameras were army surplus. Various bits and pieces were donated by sympathetic scientists at other London colleges. Their DNA crystals came from the thyroid gland of a calf in Switzerland, and when they suspended each sample in hydrogen to get a sharper image they had to use paper clips.

Everything was pretty primitive. The X-ray images had to be captured at night, when the laboratories were empty because King's College was afraid hydrogen might leak from the lab. The college's concern was justified; hydrogen leaked from the edges of the apparatus that contained the DNA, so Wilkins and his colleagues sealed samples by sheathing them with a condom.

Somehow, in spite of working in such primitive conditions and with exposure times of five hours or more, Maurice Wilkins managed to produce fascinating and highly revealing images. By early 1951, Wilkins had produced an X-ray image that suggested DNA had a regular structure. The story goes that he took a break one night and stepped outside his lab for a breath of fresh air. Ahead of him, on the far side of the river, was the OXO tower. Something about the 'X' in the bright red OXO sign triggered a thought: the structure must look like a helix.

Wilkins' insight was on the right track, but a few months later worrying news came from California. Linus Pauling, the *éminence grise* of molecular architecture and doyen of X-ray crystallographers, had published a series of papers. They showed that many molecules of fibrous proteins similar to DNA were shaped like corkscrews – helical structures, in scientific terms.

Gradually the pieces were coming together. It was clear that the scientific world was on the verge of discovering the structure of DNA and the most likely person to discover it would be Pauling, who everyone suspected would now focus his attention entirely on the mysterious molecule.

At King's College on the Strand in London, Wilkins was making good progress, but for all his talent and ability he was a self-effacing and reserved man who, his critics said, lacked a sense of urgency. Maybe this was because Wilkins had already achieved enough not to have the burning ambition of younger scientists. However, some historians have suggested it was because of Wilkins' attitude towards a bright and highly talented young research assistant in his department called Rosalind Franklin.

By her early thirties Franklin had built up a formidable reputation as one of the best X-ray crystallographers of her generation. Having accepted an offer to move from Paris to King's College, she was under the impression that she alone at King's would have the task of applying X-ray science to the study of DNA crystals.

Franklin arrived at King's while Wilkins was on holiday. On his return, Wilkins at first thought she had been hired as his junior assistant. That was an insulting blunder. Then he thought her appointment meant that he had been sidelined, so he complained. Another big mistake. Their relationship never recovered. From then on, Franklin and Wilkins bickered and fought. It was a classic personality clash: Franklin's direct, precise, decisive logical manner against Wilkins' shy, passive, speculative ruminations.

Franklin was critical of most of the work that had been done on DNA thus far. She was doubtful about Wilkins' helix, considering his results inconclusive and being of the opinion that he had not done enough scientific research to prove his theory. And she had no time for molecular models made of bits of metal plate and wire. When she visited Watson and Crick in Cambridge towards the end of 1951, she practically laughed out loud at their latest 3D visualisation. It was all wrong, she said. They had seriously underestimated the water content of DNA, which meant their structure was distorted. She gave it to them straight: 'You've got it inside out!'

Despondent, Crick returned to his PhD thesis (he had already twice been warned off his DNA obsession by his supervisors). Watson was equally shaken by Franklin's comments; he portrayed her later in his book as difficult and secretive. There was some truth in that comment – Franklin had an upper-class haughtiness – but, as we shall see, she also had very good reason not to trust her colleagues and rivals.

However, Watson's dislike of Franklin went deep, as further comments in the book reveal. Describing her as 'not unattractive' – indeed, a woman who 'might have been quite stunning had she taken even a mild interest in clothes' – he accused her of dressing with 'all the imagination of English blue-stocking adolescents'. And as if such jaundiced observations were not enough, he went on to write, 'clearly Rosy had to go or be put in her place', adding: 'the thought could not be avoided that the best home for a feminist was in another person's lab'. Only Watson knew quite why he needed to make such snide comments about Franklin, but ultimately, if anyone had cause for resentment, it was Franklin.

The truth is that Franklin was quietly producing by far the clearest and best X-ray diffraction photographs of DNA that anyone had ever seen. To this day they are considered to be among the most beautiful X-ray photographs of any substance ever taken. From these it was possible to work out the precise positions of all the fragments that make up the mammoth molecule. She had discovered that it exists in two forms, called A and B, and that the 'backbone' – comprised of ribose sugar and phosphate – lies on the outside of the molecule, unlike in Watson and Crick's model, which she had so roundly ridiculed, which had the bases on the outside. In spite of her passionate resistance to the helix idea, Franklin's notebooks suggest that she continued to think about it, wondering whether there could be a two-chain version, a double helix.

Franklin kept these results close to her chest. Frequently patronised by Wilkins and other male members of the department, and, like all female researchers at King's in the 1950s, ostracised to the extent that she was not allowed to eat lunch in the same common room as her male colleagues, it is perhaps not surprising that Franklin refused Wilkins access to her work or results. Some have suggested she was an egomaniac, others that she was simply trying to make her way in a male-dominated workplace, but whatever the truth, her efforts to conceal her work proved unsuccessful, for Wilkins surreptitiously gained access to her X-ray results, including to the image that would change the world.

Photo 51, as it is known, was taken by Rosalind Franklin and her PhD student Raymond Gosling on 2 May 1952. Franklin immediately recognised its implications, but for months this crucial image just lay in a drawer.

Meanwhile in Cambridge, Watson and Crick had almost given up. Then a succession of fortuitous events reawakened their interest.

First, a mathematician colleague, John Griffith, worked out from the molecular shapes of the bases that adenine and thymine could fit together like two pieces in a jigsaw. He found the same applied to cytosine and guanine, and that any other combination of bases was highly unlikely, if not impossible. Then Chargraff paid a visit to the Cavendish laboratory and reminded them of his earlier research. Neither of these was enough in itself to motivate

Watson and Crick back into action, but they did start thinking about the DNA conundrum again. Finally, they heard news that Linus Pauling had solved the structure. But when in early 1953 they examined Pauling's findings, they were convinced he had got it wrong. He had missed out the acid component. They realised equally, however, that Pauling would soon spot his error and come to the correct conclusion. If Watson and Crick wanted to fulfil their ambition, they needed to act fast.

Watson immediately went down to London to consult with Wilkins. Although relations between the Watson–Crick team and that of Franklin were frosty, Wilkins had maintained a good rapport with his Cambridge counterparts. Watson and Crick had lent Wilkins some workshop jigs so that his team could fabricate atomic components to build the molecular models that Franklin so disdained. Wilkins told Watson he had something to show him. It was Photo 51, to which Wilkins had recently gained access after Franklin had moved to the physics department at Birkbeck College.

'The instant I saw the picture my mouth fell open and my pulse began to race,' Watson later wrote. On the train back to Cambridge, he sketched on his newspaper the details he remembered from the picture, including clues to the angles and spacing within the helix. By the time he was home he had decided that its most likely structure was a double helix, and that he and Crick should immediately build a model to see if the pieces would fit.

For the next month, Watson and Crick built model after model, trying to work out the three-dimensional shape for DNA that would match Franklin's X-ray image. Fearing that Pauling would beat them to it (as Watson put it, he and Crick were 'scared to death'), they worked at such a furious rate that the machine shop which supplied them with the tin bases for their models could not keep up. They had to resort to strips of stiff cardboard to represent the strands of DNA.

The breakthrough came in February 1953. Franklin's X-ray image had confirmed that DNA was a helix, but now Watson and Crick realised there was more in Photo 51 and other X-ray images than they had first noticed. The images showed crosses; each arm of the cross had a series of distinct marks like rungs on a ladder. They realised that the fourth rung was always missing. Watson suddenly understood that the helix must have two strands, which were crossing over at just that point. That was why the fourth rung was missing. It meant that DNA must have a double helix: two chains wound round each other. Crick agreed, and everything seemed to fall into place. They made another model. At last it all made sense.

DNA *was* a double helix: two corkscrews coiled around each other. And the sequence of bases connected the two helices like the rungs on a spiral staircase. Cytosine to guanine; adenine to thymine. It was beautiful and elegant. And best of all, it worked.

The order of the bases provided a blueprint – a gene – for the proteins that are the building blocks of life. Thus the sequences of bases could spell out a code to make proteins in the same way that sequences of letters in the alphabet can spell out words, sentences, poetry and prose. By untwining its two helices and copying each of them, the DNA molecule could pass on that code to offspring.

Watson and Crick had done it. They had beaten everyone else. On 28 February 1953, they rushed into The Eagle pub in Cambridge, Crick shouting to anyone who would listen that they had found 'the secret of life'. The structure of DNA had been found.

> Yes, Watson and Crick had a lot of luck, but let's not take away from their achievement. As Crick himself said, 'It's true that by blundering about we stumbled on gold, but we were looking for gold.'
>
> And the biggest nugget of all? Forget all mystical ideas of an essence of life or a life force. From now on, the essence of life was just bytes and bytes of digital information in the form of the four bases adenine, cytosine, guanine and thymine.
>
> Richard Dawkins

A few weeks later, in the 25 April 1953 issue of *Nature*, the world's most prestigious scientific journal, Watson and Crick announced their model of DNA without making any mention of the contribution from Franklin's X-ray image. Next to their announcement was a paper by Franklin which, coincidentally, she had been working on at the time of the two men's breakthrough. It presented convincing X-ray data to support the double helix structure. A third paper by Wilkins suggested a helical structure for DNA.

The discovery of DNA's structure would transform the world of medicine, even the world of crime detection, and lead to an explosion in the fields of biochemistry and genetics. Nevertheless, nine years later, when the Nobel Institute decided to award a Nobel Prize in Physiology or Medicine for the discovery of the structure of DNA, it went to Watson, Crick and Wilkins. By then, Franklin had died of ovarian cancer. Nobel prizes are never awarded posthumously, but they are also never shared between more than three people.

* * *

The mapping of the architecture of DNA by Watson, Crick, Franklin and Wilkins led to a slew of further genetic break-throughs by British scientists, most of them with momentous and significant implications for the future.

In 1962, John Gurdon, an Old Etonian molecular biologist working at Oxford University, announced that he had cloned a South African clawed frog. He had used ultraviolet light to destroy the nucleus in a frog egg, in effect creating a hollowed-out egg, into which he had then implanted the nucleus of an adult frog's intestinal cell. The tadpole that resulted was genetically identical to the adult frog from which the intestinal cell had been extracted. There were some early problems – the first generations of cloned tadpoles did not survive to adulthood – but Gurdon improved his methods and eventually produced sexually mature adult cloned frogs.

The principle of Gurdon's nuclear transfer cloning method is still used today. Most notably it was employed by Ian Wilmut, an English former farmhand who became an embryologist and made headlines around the world when he supervised the creation of Dolly the sheep, the first cloned mammal derived from adult cells.

Watson was later involved in the most ambitious attempt of all to apply scientific rationality to explaining human nature – the sequencing of the human genome. As founder of the Human Genome Project, Watson persuaded John Sulston, a British worm researcher who had previously unravelled the relatively puny genome of a tiny nematode, to spearhead Britain's contribution to the project. As director of the Sanger Centre in Cambridgeshire, Sulston's greatest moment came when he fought off attempts by American biotech pioneers who wanted to turn this exploration of our fundamental make-up into a multi-billion-dollar private enterprise.

Then in 2000 Bill Clinton and Tony Blair announced that the work by Sulston's team was complete. The three billion units of DNA code that determine our intelligence, stature, temperament and vulnerability to disease had been mapped, providing for the first time a definition in scientific terms of what it means to be human. The sequencing of the human genome made all kinds of new medical procedures possible, about some of which scientists such as Harvey, Jenner and even Alexander Fleming could only have dreamt.

Solving the structure of DNA in the 1950s had also set scientists, philosophers, social scientists and many others to work interpreting the implications of the discovery, not least for Darwinism. Among them was William Hamilton, a reserved and eccentric Oxford don who was less interested in how genes are constructed than in how their mixing between generations could shape the world. In attempting to answer the big questions of our existence, Hamilton devised the selfish gene theory, later popularised by Richard Dawkins, and in the process gained a reputation as the greatest biologist of the twentieth century.

I would have given my eye teeth to meet Charles Darwin and to discuss with him some of the trickier implications of his theory of evolution. But I can console myself with the thought that I did meet, and get to know, perhaps the nearest equivalent that the late twentieth century had to offer. His name was Bill Hamilton – and for the better part of forty years he was, variously, my inspiration, my colleague and my friend.

Bill was a shy and forgetful man, in equal parts very deep thinker and very knowledgeable naturalist, rather like Darwin but with added mathematics. He thought all the time, seeking biological answers to big questions. Why are we animals altruistic? Why do we have sex? Why do we grow old?

Bill discovered the key to self-sacrifice, or altruism, amongst animals and insects. On the basis of his study of bees and herding animals, he explained to us that altruism was not some freak behaviour that contradicted the survival of the fittest, but an intrinsic, self-perpetuating law of evolution.

Richard Dawkins

Hamilton was never happier than when locked away in his study. Intrigued by what genes do and how they interact, he wanted to find biological answers to big questions, such as why there is sex and why we grow old.

However, the question that fascinated him more than any other was altruism. From the viewpoint of Darwin's theory of evolution by natural selection, acting in the interests of other members of the species simply did not make sense. Altruism thus appeared to throw the entire theory of Darwinism into doubt. Darwin himself had been concerned when he observed the extremes of cooperation seen among social insects, such as termites, bees and ants. Evolutionary biology saw animal behaviour as always devoted to the good of the species, and in that context altruism – behaviour devoted to the good of another individual – seemed hard to explain. As for suggesting that altruism was a learned behaviour with a higher evolutionary purpose, it simply did not make sense to Darwinists. Why should creatures with no powers of reasoning, such as bees for instance, sacrifice themselves for the good of others? It was a fascinating area of enquiry and Hamilton was determined to discover the answer.

Hamilton grew up at Badger's Mount, a small village in Kent now just inside the M25 motorway. For him it was paradise. He could chase butterflies across the surrounding fields and roam barefoot down the country lanes. He would rip the bark off dead branches to reveal what he called the funeral feast of trees: woodlice and other insects grubbing for food. He found beetles in the outhouses and birds' nests in the trees. It was a perfect environment for a growing naturalist.

Although he might have been shy, Hamilton certainly was not timid. As a child he had an all-too-close encounter with some explosives that his father had left lying around after trying to make home-made grenades for the Home Guard. The young Hamilton experimented with these in a shed at the end of his parents' garden, trying to make a small bomb, but suddenly some of the highly combustible materials went off in his hands. Larking about with explosive cost him the tips of two of his fingers and left a piece of shrapnel lodged in his lung.

The Meaning of Life

This most respected of Darwinians had almost met his maker before he'd done his work, but at least the accident saved him from doing anything especially active during his National Service in Hampshire. Instead Hamilton spent most of his time at a desk or visiting his aunt, who lived nearby and was an expert on beetle collecting.

As an adult, Hamilton took more care to ensure that the only edifices he blew up were established notions. In their place he erected ideas that were arguably stranger, more original and more profound than those of any other biologist since Darwin.

Richard Dawkins

Insects held a particular fascination for Hamilton. On several occasions Hamilton said he preferred the company of bugs to human beings. As a young man he was tall and good-looking, but that never gave him the confidence he needed to overcome his shyness, which probably explains why he thought insects made for better company. In fact, when he moved into a bedsit near Windsor Park, he was delighted to find it was infested with minute bacon beetles. He treated them like pets.

From an early age, Hamilton realised that the behaviour of insects can tell us a lot about animal instinct. One day he watched a honeybee worker commit suicide in an attempt to save her colony. When she stung a marauder of her nest, she was unable to withdraw her sting, so she flew away, leaving behind her sting and a portion of abdomen. A few minutes later she was dead, yet her poison gland was still pumping venom into her victim. Why, Hamilton wondered, would a bee go so far as to die to save her hive?

As a postgraduate, however, without any financial backing, Hamilton had to rely on his own resourcefulness to answer such questions that puzzled him. He became in effect an itinerant scholar, picking up bits of work wherever he could get them. Dividing his time between the London School of Economics and University College London, he was ignored at one institution and mistrusted at the other. It was an extremely lonely life, living in a bedsit in Chiswick and working in libraries. Sometimes he would take a pile of books to Waterloo Station just to be surrounded by the commuter throng.

For much of his life, it must have seemed to Hamilton that the world was against him. No one was willing to support altruism or genetics, or for that matter Hamilton himself. But this was the early 1960s and not long after World War Two. People were nervous about anything that smacked of the Nazi interest in eugenics. It was one thing for Watson and Crick to discover the structure of DNA, quite another to suggest that human behaviour, all human traits and skills, might be determined by it. That might give the idea that some people were genetically superior to others.

Then in 1964 Hamilton took off. In the great tradition of British naturalists such as Russell Wallace and Joseph Banks, he set out on an expedition to the Amazon, where he planned to observe social insects such as bees, wasps and termites, convinced that their behaviour would reveal the genetic clue to altruism.

On one occasion, as they were travelling upriver, the boat sprang a leak. Without a moment's hesitation, Hamilton jumped overboard and swam underneath to plug the hole until something more permanent could be done. While demonstrating his own form of altruism, he commented afterwards that the ferocity of piranhas was much exaggerated – perhaps their gnawing was less painful than losing fingertips in a bomb blast or maybe he was just keen to display his own selfless nature.

The field trip to the Amazon was a turning point. It provided Hamilton with the inspiration and evidence he needed for his major new theory. Back in Britain, he dashed off two landmark papers published in 1964 in which he questioned how it was possible that self-interested individuals could shake off their egotism to help others in a selfish Darwinian world. His answer was that selfish genes make altruistic individuals: shared genes are the key to cooperation. The more an individual shares its genes with the group, the greater the likelihood that it will be altruistic.

Leafcutter ants are native to South America. They forage for leaves, which they cut and carry home. They don't actually eat the leaves. They feed them to a fungus that they grow underground. It's the mushrooms that they eat and feed to the queen, the larvae and the other workers.

From a Darwinian point of view the behaviour of leafcutter ants could be called an act of altruism. The workers are working for the good of the reproduction of another individual, the queen. In a Darwinian world, that's a supreme act of self-sacrifice.

It was a seminal and highly important addition to Darwinian theory in the 1960s. Today it is hugely influential in my field of evolutionary biology.

The theory made sense not only of ants working to feed other ants in the colony but also of animals working to rear the young of other family members. What Bill had demonstrated was that kindness and mutual support had a part to play in natural selection.

Richard Dawkins

Hamilton expressed his theory in mathematical terms that came to be known as Hamilton's rule. It was encapsulated in the formula k>1/r. Put crudely, this says that an individual can die for its close kin and still spread its genes because the close kin will have many genes in common with it. The altruist's genes will still survive and that's what matters. Hamilton's rule produced a measure of the likelihood of an individual performing an act of self-destructive altruism. He said it depends on three factors: the cost to the altruist, the benefit to the altruist, and a coefficient of relatedness between the altruist and the beneficiary.

When, however, Hamilton submitted to *Nature* his paper, 'The Genetical Evolution of Social Behaviour', the esteemed science journal rejected it out of hand. With dense mathematics and denser prose, it was a difficult read even for the most scientifically literate people. 'I received the editor's decision almost by return of post,' Hamilton said later. 'In about three lines he regretted that he had no space for my manuscript and suggested that it might be more appropriate to a psychological or sociological journal.'

Eventually published in two parts in 1964 in the comparatively obscure *Journal of Theoretical Biology*, Hamilton's paper was the first clear statement of what has since become known as the gene's-eye view of evolution, the selfish gene school, or socio-biology – the theory that sees individuals as vehicles for competing genes.

Hamilton's work is a natural and inevitable extension of Darwinian logic.

The theory made sense of bees dying to save the hive and of animals helping to rear the young of other members of the group, and if the benefit to the group is high enough, he argued, the genes will be spread around.

Vitally important as Bill's theory was, I was amazed how infrequently his work was mentioned in books and articles on the subject. Academics just didn't seem to notice that Bill was there. Eventually that changed and the academic world woke up to Bill Hamilton. The honours and the prizes began, at last, to roll in.

Bill spent his later years as a colleague of mine at New College in Oxford. We were good friends. The extraordinary thing about him was that he sometimes didn't recognise his own genius. He'd talk about the ideas other scientists had written about and I'd have to point out to him: 'Bill, you had that idea before anyone else.'

'Did I?' he'd say modestly. 'Oh, perhaps I did.'

Richard Dawkins

Hamilton had a reputation for being slightly reckless and very adventurous. Residents of Oxford would see him cycling around the city at hair-raising speeds, often winding up in an accident with some astonished motorist. Looking for ants on one of his expeditions, he hiked through Rwanda at the height of the civil war there. Unsurprisingly, he was treated as a spy, so implausible was his explanation for his escapade. On another occasion he was held up at knife-point in Brazil and made the mistake of fighting back. He was viciously wounded.

Hamilton's very last expedition, to the Congo in 2000 to collect the faeces of chimpanzees, was to end in tragedy. He had gone in the hope of finding evidence in the chimps' faeces that HIV originated in an experimental polio vaccine tested in central Africa in the 1950s, a theory that was widely discredited by most scientists. Two samples tested positive, but the news came too late for Hamilton. While travelling, Bill caught malaria. As well as antimalarials, he took soluble aspirin, not knowing or maybe not remembering that he was suffering from ulcers. The aspirin caused a massive stomach haemorrhage. Hamilton fell into a coma and died. It was a dreadfully sudden and premature end for such a great talent.

Bill once said that he hoped that when he died he would be laid out, unburied on the floor of the Amazon jungle and interred by burying beetles. In fact, he's buried in the cemetery at Wolvercote on the edge of another famous forest, Wytham Wood, near Oxford.

The Meaning of Life

> At the graveside, his devoted companion Luisa brought us to tears as she invoked one of Bill's wilder theories. Luisa conjured up a wonderful image of Bill's body, broken down by bacteria and transformed by the fungi spores, being carried up into the clouds, from where he might eventually rain down on the forests of his beloved Amazon.
>
> Richard Dawkins

* * *

While Crick and Watson were working to unravel the mysteries of life, half a mile from where they worked another remarkable Englishman was absorbed in equally deep research. Instead of looking to discover what makes each of us so unique as individuals, Fred Hoyle was searching for something far more fundamental: the answer to where we all come from. In fact, he wanted to discover how *everything* came to exist – all the atoms of all the elements in the universe.

Fred Hoyle is remembered for three things. To some he is the person who produced a bunch of very successful science-fiction novels and television series, such as *A for Andromeda*. Others will recall him as being the first to coin the term Big Bang, although he used it contemptuously to ridicule the idea that the universe had expanded from a primordial hot and dense initial condition at some finite time in the past and continues to expand to this day. Instead, Hoyle believed in an alternative theory – one that he had developed with other scientists – which pictured the universe in a steady state that has always looked the same as it does now. Under the steady state model, the universe is eternal, constantly replenishing itself to fill the spaces that arise as galaxies expand, then fall apart. We now know that Hoyle was wrong. All the evidence points towards the Big Bang theory being correct. However, in the course of his research into the steady state theory, Hoyle made his third and greatest discovery, namely the method by which carbon and all the other elements in the universe are created from hydrogen.

Today's understanding of the building of chemical elements from the lightest, hydrogen, to the heaviest, uranium, during the lifetime of stars will always be linked to Fred Hoyle's fundamental work. And of all the elements, it is the formation of carbon

307

that is particularly important. Carbon has unique chemical and physical properties that allow it to form the complex organic molecules that create and sustain life. Without carbon there would be no living things. No plants, no bacteria, no animals and no us. In other words, carbon is essential, and Hoyle's discovery of how it originates was a true Eureka moment that for a quarter of a century turned Fred Hoyle into the most famous astrophysicist in the world.

Like Watson and Crick, Hoyle was an unlikely candidate to become a scientist at Cambridge. Born in 1915 in the Yorkshire village of Gilstead, he grew up in a valley dominated by the mills of the textile industry. However, Gilstead was on the edge of Crockley Moor, allowing Hoyle to lead the sort of wild outdoor life that he much preferred to being stuck indoors. In fact, Hoyle was as far removed from the geeky stereotype of a scientist as anyone could be. Frequently playing truant from school (partly to escape beatings for challenging the teacher's knowledge), he was always off on the moor or playing around the canal near his home. All his life Hoyle loved the outdoors, whether hiking or mountaineering. He even courted his future wife – with whom he was to enjoy a particularly long and happy marriage – on a hiking trip.

> Hoyle used to say about the hours he spent mucking about on the canal that he was instinctively displaying a sound sense of engineering. Watching lock gates and sluices open and close was much more valuable, he thought, than anything he could have learnt at school.
>
> Jim Al-Khalili

However, there was another side to Fred Hoyle: insatiable curiosity. At the age of 10 he found some chemistry books on his parents' shelves. After trying some elementary experiments, he decided to make some phosphine gas.

'Does tha 'ave any glass tubing?' he demanded of the local pharmacist. 'Any concentrated sulphuric acid? H_2SO_4, tha knaws.'

But it was not only his curiosity that put young Hoyle on the road to international fame. His mother, a talented pianist who played Beethoven sonatas to accompany silent films at the local cinema, had ambitions for her son. She ensured that Hoyle did well enough at primary school to win a scholarship to Bingley

Grammar School and then a Yorkshire scholarship to Emmanuel College, Cambridge, where he won prizes for an essay on beta decay and his performance in the mathematical Tripos.

Top of his year in theoretical physics, Hoyle also gained a reputation for his self-confidence – some people called him cocky – and throughout his time at Cambridge he was known for his belligerence and speaking his mind, a constant thorn in the side of the British scientific establishment.

Hoyle appeared set for a doctorate and a secure job at the university when the Second World War intervened. By the end of 1939 he was married to Barbara Clark and the war effort had scooped him up to work on the development of radar. With every scientist doing their bit, Hoyle had no choice, but he was frustrated not to be able to give more time to his new interest in stars and how they developed.

Very little was known about the universe in the 1940s. In 1929 the American astronomer Edwin Hubble had provided proof that the universe was expanding. He also suggested a date for creation: 1.8 billion years ago. In Fred Hoyle it prompted a desire to discover how the whole thing began. Fortunately, he found himself working alongside two Jewish émigrés from Europe who would play a vital part in his research on the universe – Thomas Gold and Hermann Bondi. From 1945 the three of them took up posts in Cambridge, and, together, they struggled to apply Einstein's ideas about relativity to cosmology. At the heart of their investigations was a single question: how did the universe get started?

The trio would meet in Bondi's rooms at Trinity College. Bondi, the mathematician, sat on the floor, while Hoyle hurled ideas at him to turn into breathtaking mathematical equations. Progress was slow. But unlike with most scientific discoveries, which are usually the result of painstaking incremental advances in understanding, the trio's breakthrough came as the result of a sudden Eureka moment of inspiration. It came from a quite unexpected source.

Thomas Gold had been to the cinema to see Michael Redgrave in *Dead of Night*, a film that contained a recurring nightmare. When Gold next saw Hoyle and Bondi, he made a remark that got them thinking. 'Suppose the universe were like that,' he said, 'with no beginning and no end, constantly re-cycling?'

Seizing on the idea, Hoyle and Bondi made it the basis of their brand new theory of the universe, which came to be known as the steady state theory or the continuous creation theory. As Hoyle later described it, the idea came to him over the course of several feverish hours, working in his room late one night. 'I think it was on the tenth of February, 1948, that one evening I got down and really got the equations on paper. I remember I had them somewhere about ten o'clock in the evening. I sat up until about three in the morning and by then I had the solution that I wanted.'

The equations suggested that the universe did not begin in a single moment, the Big Bang, but had always existed. As this universe expanded, galaxies would speed away from us, leaving gaps that were filled with new matter, which was constantly being created.

These were revolutionary, controversial ideas and Hoyle was keen to let the whole world know about it. Hoyle's reputation soon extended well beyond the academic world, enabling him to establish a very successful second career in broadcasting. Not everyone liked his Yorkshire accent, but his talks were hugely popular. One producer remarked that he tackled events in interstellar space as if he were commenting on a cricket match.

The Fred Hoyle Collection, held at his Cambridge college, St John's, is fascinating. It contains his first telescope, his walking boots, his crampons and even his backpack, complete with glucose tablets.

It also houses a carbon copy of a radio talk that Hoyle gave in March 1949, in which he invented the term Big Bang. He used it derogatorily. Hoyle thought the Big Bang was like a girl jumping out of a birthday cake. It was a party trick, he said, not a serious theory. Here's what he said about it:

'It is an irrational process that cannot be described in scientific terms. On philosophical grounds too, I cannot see any good reason for preferring the big bang idea. Indeed it seems to me in the philosophical sense to be a distinctly unsatisfactory notion since it puts the basic assumption out of sight where it can never be challenged by direct appeal to observation.'

Jim al-Khalili

However, other scientists took the Big Bang theory seriously, and during the 1950s Fred Hoyle's steady state theory of the universe was overtaken by the growing amount of evidence in support of the Big Bang. During the course of his research, however, Hoyle had come up with another theory that he called nucleosynthesis, which would have a much longer lasting impact on cosmology. For years, scientists had speculated about how carbon – the element that is essential to all organic compounds and therefore critical to all life and to us – was created. Hoyle found the answer deep within the nuclear explosions that take place inside stars.

Ordinary stars like the Sun derive more than three-quarters of their enormous energy through fusing hydrogen atoms, which under enormous gravitational pressure combine to form helium and convert the excess mass into energy in accordance with Einstein's theory of relativity. About 15 per cent of the energy produced by the Sun comes from more complex fusion reactions resulting in lithium and beryllium. The combination of all these fusion reactions accounts for the creation of the first four elements in the periodic table, prompting Hoyle to wonder how the remaining 78 stable elements were formed.

Hoyle realised that the enormous heat and gravitational force within a star would be sufficient for three helium nuclei (or alpha particles) to collide and survive for long enough to form carbon. Then, in a series of chain reactions, successive alpha particles would fuse with carbon to form oxygen, neon and so on, until we reach iron, the 26th element in the periodic table.

Generating elements beyond iron requires the extraordinary conditions generated when a star implodes into a supernova. In temperatures of more than 10 billion Celsius, the heavier elements from iron to lead are formed, including most metals such as copper, zinc, gold, silver and platinum. Supernovas last only a short time, then explode, upon which they scatter into the universe the elements formed in their last dying moments.

In a very different sense from Watson and Crick, Hoyle had also found the secret of life. The elements that make up all our bodies were formed in the stellar or supernova fusion reactions he had identified. And remarkably, evidence for his theory was immediately found. The proportion of elements predicted by Hoyle's theory of nucleosynthesis proved to be very close to the distribution actually found in the natural world. Despite this, Hoyle's nucleosynthesis theory was so far ahead of its time that it was widely ignored for about ten years.

Some years later, at a lecture to the Royal Society, Hoyle was talking about his theories when a young PhD student stood up. His name was Stephen Hawking and he pointed out that Hoyle's ideas about carbon would only work in the wake of a Big Bang. Hoyle had to concede that Hawking had a point, even though he considered that Hawking had behaved unethically by speaking out in front of the Royal Society.

Despite his having breached Royal Society etiquette, Hawking was in fact a great admirer of Hoyle. He had come to Cambridge specifically to study under him and certainly had not intended to humiliate the great man. He was simply a young, talented student casting around for something worthwhile to study and not at all sure what that might be. His talent would later take him as far as the Lucasian Chair of Mathematics, the post that Newton had occupied 300 years earlier, and it would make him the most famous scientist since Einstein. But if anyone had forecast such fame for Hawking when he graduated in 1962, let alone suggested that he would write a book on relativity and black holes that would sell over 20 million copies, no one would have been more dismissive of the idea than the then 20-year-old Hawking.

Hawking did not learn to read until he was eight. At school he was a very average student, not even making it into the top half of his class. At Oxford he enrolled at University College intending to study mathematics, but when he arrived he discovered that mathematics was not offered at the college, so he chose physics instead, given that this suited his interests in thermodynamics, relativity and quantum mechanics. Hawking's innate talent was immediately noticed, his physics tutor noting that 'it was only necessary for him to know that something could be done, and he could do it without looking to see how other people did it. He didn't have very many books, and he didn't take notes. His mind was completely different from all of his contemporaries.'

However, Hawking's genius was not matched by his study habits. His final examination score sat on the borderline between first and second class honours, making an oral *viva* examination necessary, after which his tutor said, 'The examiners were intelligent enough to realise they were talking to someone far more clever than most of themselves.'

After a few postgraduate months spent studying astronomy, Hawking realised he was much more interested in theory than observation. There being no one working in cosmology at Oxford at the time, Hawking went to Cambridge, attracted by Hoyle's reputation.

At about this time, Hawking began to notice that his body was not working properly. He had become clumsy and his speech was slurred. He suspected it was serious. This was the beginning of the motor neurone disease that was to take over his life. Over the next forty years or so, his family, friends and colleagues would watch as he slowly lost all neuromuscular control of his body. But as his body wasted away, his mind soared.

On arrival in Cambridge, Hawking found that he was not to have Fred Hoyle as his supervisor after all. It was a disappointment because, like Hoyle, he wanted to use mathematics to investigate the origins of the universe. However, a stroke of luck kept him in touch with Hoyle's ideas. Hawking found himself working in the same lab as one of Hoyle's PhD students. Since he had been a boy, Hawking had believed in the idea of an unchanging and everlasting universe – the steady state model – but when he caught a glimpse of Hoyle's equations he began to suspect he was wrong and, what's more, that Hoyle was wrong too.

Twentieth-century physics was dominated by two equally important areas of study. One was general relativity: Einstein's theory of gravity describes the stars, the galaxies, the whole Universe – in other words, the very large. Quantum mechanics, the second area of study, deals with the infinitesimally small: the subatomic world. Hawking realised that the secrets of the universe must involve these conflicting theories working at the same time.

In the 1960s, black holes were cosmology's hot topic. Einstein's theories of relativity, published in the first two decades of the twentieth century, predicted that stars could collapse to form black holes. According to Einstein, they were points in the universe where space was sucked in on itself.

Now, fifty years later, the work of Roger Penrose, a mathematician who had recently been awarded his doctorate at

Cambridge, caught Hawking's eye. In particular, his work on the singularity at the centre of black holes fired up Hawking's imagination. A singularity is a place of such infinite gravity that it appears to suck in everything – matter, light, space and time – as if they are all merely water disappearing down a plug hole. With every new discovery black holes were becoming more mysterious, and Hawking's work was about to add to their enigma.

> Working extraordinary hours with Penrose, becoming restless and agitated by the difficult problems he'd set himself, Hawking came to the realisation that at the point of the singularity the general laws of physics must break down.
>
> Their research produced some truly startling results about the origin of the universe, including that black holes aren't all that black.
>
> Busy as they are, sucking in anything and everything that comes near, black holes also spit out an almost invisible, glowing shower of subatomic particles, or what we now call Hawking Radiation.
>
> If an astronaut fell into a Black Hole, he wouldn't reappear but his mass-energy would be thrown out again as radiation.
>
> Jim al-Khalili

Hawking's discussions with Roger Penrose revealed another even more astonishing insight: the universe itself was born in a singularity. It suggested the Big Bang theory was actually true. At the moment that the universe began, black hole behaviour reversed. Instead of everything being sucked in by a black hole, matter, time, space and energy all suddenly exploded from the singularity at the heart of a black hole. Hawking and Penrose's explanation meant that Fred Hoyle's steady state theory of the universe was wrong, and their own theory has since become the established model for the creation of the universe.

Hawking's work continues. He is still searching for a theory of everything: the explanation that will tell us not only how the universe works, but why it began. It is the most fundamental question of all – not how or when did we begin, but *why* we are here.

There is a very direct link between the search for truth by Stephen Hawking and the same search by William Gilbert, the Elizabethan doctor whose investigations of magnetism started the ball rolling in 1600. They are linked by a quest for a rational and objective truth to describe our existence – a truth tested by experimental investigation and which cuts through the mysticism, conjecture and superstition that dominated until such geniuses of Britain as Newton, Hooke, Halley, Darwin, Faraday, Maxwell, Rutherford, Dirac, Watson, Crick and Hawking brought science into our lives.

Thanks to these British scientists, British science now stands on the cutting edge of new technology and on the threshold of new breakthroughs. Already we are catching a glimpse of the kind of discoveries and inventions that scientists will make in the future in fields such as genetic engineering and cosmology. Among the most exciting are the materials science breakthroughs of nanotechnology, which has its origins in the discovery by Harry Kroto, a British chemist, of a new type of carbon molecule with sixty carbon atoms. The discovery, which won him the Nobel Prize in Chemistry, was the inspiration for carbon nanotubes, which are incredibly tough but also extremely light – one gram of nanotube stretches for twelve miles. It is the most revolutionary and versatile material ever discovered. It has vast applications, ranging from a highly conductive yet extremely light replacement for copper wiring to a cable that would extend 22,000 miles into space to connect to a satellite – a space elevator that could convey people and cargo into space.

The remarkable journey that has brought us to the world we live in now, and which will lead on to the discoveries of the future, was taken by scientists who were often physically brave and always intellectually fearless. In the Middle Ages and in the early days of the scientific revolution, many scientists risked their lives by developing and publicising 'heretical' ideas that threatened the Church's creationist dogma. William Harvey knew his investigation of the role of blood vessels and the heart could result in the same fate as Michael Servetus, a Spanish physician, burned at the stake for committing alleged heresies during his publication of his findings concerning the pulmonary circulatory system.

The list of men – until the late nineteenth century, the upper echelon of British science was sadly closed to women – who risked their health, lives, reputations and freedom to discover the answers to fundamental questions about the world, the universe and our place within them is long and illustrious.

Newton stuck a needle into the socket around his eye in the hope of discovering how his sight worked, a quest that led to his work on optics and eventually to his considerations of gravity and the fundamental forces that bind our universe. Halley ventured across some of the roughest seas of the South Atlantic to make the astronomical observations that established his reputation, while Cavendish subjected himself to electric shocks to investigate the mysterious electricity and lost his eyebrows to the explosive effects of hydrogen.

Infecting a young boy with smallpox might have cost Jenner his reputation and his liberty if he had been wrong, but he thought his audacious procedure was a worthwhile risk if it validated a vaccination theory that in fact saved the lives of billions. Davy subjected his lungs to a succession of previously unidentified gases and paid the price with an early death. Ray, Willughby, Banks, Lyell, Darwin and Wallace travelled to some of the most inaccessible and potentially dangerous parts of the planet in a spirit of exploration that eventually led to the discovery of the origin of species by evolution and natural selection.

As well as a healthy disregard for their own well-being, these great men of British science frequently share one other characteristic. Few of them came from a conventional background or underwent a typical education. If Newton had not been sent away to relatives by his widowed mother he most probably would have become a grumpy and somewhat inept farmer; instead he gained a good education – much of it self-taught – and became the greatest mind of the scientific revolution.

Hooke was largely self-taught and financed his own education; Davy taught himself anatomy, chemistry, botany, physics and mechanics after his father's death; the young Darwin showed little academic aptitude and no real interest in the family tradition of medicine and science and seemed destined for a quiet life as a country vicar until he was approached out of the blue to join HMS *Beagle* as its on-board naturalist.

As one of ten children of a poor blacksmith, Faraday had a very rudimentary education and had no mathematics skills by the time he finished his apprenticeship, but with a curious mind he turned this to his advantage, becoming a supreme experimentalist who, instead of using mathematics, visualised and explained his discoveries in terms that provided a fresh perspective and which lay people could understand.

Maxwell and Dalton were precociously bright children who suffered at the hands of tutors, teachers or classmates who were suspicious of their intellect. Similarly, Turing's mathematical genius brought him into conflict with the classics- and humanities-centred curriculum of the between-the-wars public school system. And Hawking, who did not learn to read until he was eight, remained a very average student throughout his school years, not even making it into the top half of his class.

The modern world owes a great debt to these brave iconoclasts and their fearless determination to pursue their own paths of education and discovery, often clambering on to each other's shoulders in their pursuit of knowledge and understanding of our world and our existence. In 1600 the world was a mysterious place, ruled by religion and superstition, a place where we could see little further than our eyes would let us. Thanks to Hooke's work with his microscope we discovered the tiny world around us. After Newton we understood the force that kept us anchored to the ground. Later Banks opened up the rich world of plants and exotica; Hunter explained to us the workings of our own bodies.

The discoveries came thick and fast. Darwin and Wallace explained how life developed. Watt, Faraday and Maxwell gave the world power and the engineer scientists such as Brunel, Thomson and Whittle put a girdle round the Earth and gave us global communication. We learned how to picture the enormity of the universe and the tininess of atoms. We discovered how to cross oceans and fly at supersonic speeds, how to power turbines and light up cities. We unravelled fossil records and the secret codes of life. We conquered diseases, enemies and ignorance.

The genius of British scientists has kept Britain at the forefront of scientific progress for four centuries. They were often awkward and contentious characters, but all of these scientists, engineers and inventors had one thing in common: they asked the difficult questions. They spoke their minds and they were not afraid to be different. They stuck out for what they believed in. Their stories explain why science is so important to us and they show us that it is not necessary to be the cleverest student in the class or to go to the best school to be a scientist. What is needed is to be curious and fascinated by what lies around us. The world is full of wonders, but they become more wonderful – not less wonderful – when science looks at them.

Britain's geniuses changed our world. Tomorrow's scientists will no doubt change it further, and British science will be as dramatic and world-changing in the twenty-first century as in any other period in history. It is a history to be proud of – and an exciting future for all of us.

KEY DATES IN BRITISH SCIENCE

c700   While translating Greek natural philosophers into Latin, Bede the Ecclesiastical establishes a method for calculating the date of Easter. He also investigates Earth's spherical shape, the Moon's governance of tides and the shortcomings of the Julian calendar.

c870   Alfred the Great translates natural philosophy texts by Bede and Beothius from Latin to Anglo-Saxon.

c1016   The Viking king Canute establishes a tradition among English monarchs of appointing Lotharingian clerics as bishops and masters of schools; they bring new ideas in natural philosophy to England.

c1070   William the Conqueror invites several Lotharingian clerics to Britain and the West Country becomes a centre of scientific thought.

1079   Robert the Lotharingian, a skilled mathematician and advocate of the abacus, is appointed Bishop of Hereford.

1092   In the first known use of an astrolabe in the West, Walcher of Malvern calculates a set of lunar tables from an eclipse on 18 October 1092.

c1116   Working in the West Country, Petrus Alfonsi translates a complete set of astronomical tables by the Persian mathematician al-Khwarizmi, the first evidence of their existence in the Latin West.

c1125   Following a seven-year European tour, Adelard of
Bath writes *Quaestiones naturales*, then writes about
Islamic science, cosmology, the abacus, spherical
geometry, Arabic mathematics and Euclid's *Elements*.

c1141   Robert of Ketton and Robert of Chester translate
Arabic texts on algebra, geometry and alchemy into
Latin.

1167   Louis VII of France expels all foreigners from the
University of Paris. Encouraged by Henry II, the
returning scholars congregate around the monastic
schools in Oxford and begin formal lectures leading
to the formal acknowledgement of Oxford as a
university in the early thirteenth century.

c1175   Daniel of Morley publishes *Philosophia*, two volumes
dealing with man, the creation of the world, matter,
the elements, the nature of the stars and the
usefulness of astrology.

1176   Roger of Hereford writes an ecclesiastical computus
to calculate the Church calendar.

1180   Alexander Neckam lays down first Western description
of a compass in *De naturis rerum* and *De utensilibus*.

1209   The university at Oxford disperses after clashes
between scholars and townspeople, some of the
students convening in Cambridge, thereby leading to
the establishment of a university rival to Oxford.

1220   Working in Oxford, Robert Grosseteste writes widely
on natural sciences, astronomy, acoustics and optics
and argues that scientific reasoning should be verified
through experimentation.

c1242   In his nineteen-volume encyclopaedia, *De
proprietatibus rerum*, Bartholomaeus Anglicus writes
on all the branches of science and philosophy known
at that time: theology, philosophy, medicine,
astronomy, chronology, zoology, botany, geography
and mineralogy.

1249 The first Oxford colleges, which include Balliol, Merton and University College, are founded in the period leading to 1264.

1264 Often called the first scientist, Roger Bacon puts forward his idea of a science of experience. He uses experimental methods in alchemy, applies geometry to the science of lenses and undertakes experiments with gunpowder.

1284 Peterhouse, the first Cambridge college, is founded.

c1300 The Merton calculators apply mathematical or logico-mathematical methods to questions of natural philosophy, including the forces involved in the movement of objects.

1315 Lasting two years, the Great Famine brings an end to 250 years of population growth and triggers the first fundamental questioning of the institutional authority of the Church, laying the foundations for later scientific and philosophical investigations of subjects previously regarded as heretical.

1321 William of Ockham proposes the principle of parsimony, later known as Ockham's Razor. It has profound implications for the development of the scientific method.

1337 Lasting until 1453, the Hundred Years' War between English and French kings for the throne of France begins.

1348 The Black Death enters England through the port of Weymouth, killing up to half of the country's population by 1666.

c1350 Scientific exploration and thought in Britain grinds to a halt in the face of pestilence, famine and wars.

1453 Start of the Wars of the Roses, which last until 1487.

1453	The Fall of Constantinople sends Byzantine scholars fleeing to the West, bringing ancient texts that stimulate a renewed interest in logic, deduction and mathematics in the West.
1454	The invention of the movable-type printing press by Johannes Gutenberg at Mainz in Germany hastens the spread of new ideas.
1492	The discovery of the Americas by Christopher Columbus and exploration by British mariners stimulates a greater interest in geography, navigation, the compass and astronomy.
1600	The scientific revolution in Britain begins with William Gilbert's publication of *De magnete*, in which he investigates magnetism, pioneers the study of electricity and is first to advocate the scientific method that nothing can be taken for granted if it cannot be proved by extensive observations from repeatable experiments.
1603	William Gilbert dies in a plague epidemic that also kills Peter Short, the printer of *De magnete*.
1614	John Napier publishes first logarithm tables after discovering a new method of calculating using exponential numbers in 1594. He subsequently develops calculating rods and invents the decimal point.
1620	Francis Bacon publishes *Novum organum,* in which he established the concept of the experimental scientific method, thereby demolishing the Aristotlean principle of reasoning by deduction.
1622	William Oughtred invents the slide rule using Napier's logarithm scales.
1628	William Harvey publishes one of the great scientific classics, *Exercitatio anatomica de motu cordis et sanguinis*, which describes the systemic circulation of blood by the heart.

1642   The start of the English Civil War sees the university collapse after Charles I's withdrawal to Oxford, in turn sending scholars to London, where they form societies such as the Invisible College to discuss the new experimental natural philosophy of Francis Bacon.

1650   Oxford is a hotbed of scientific experimentation with 'virtuosi' polymaths including Christopher Wren, Robert Hooke, Robert Boyle and Thomas Willis meeting at the Oxford Experimental Philosophy Club.

1655   Robert Boyle employs Robert Hooke as his assistant and they embark on a period of enormous experimentation.

1660   In May, the monarchy is restored with the return of Charles II to England; in November, twelve members of the Invisible College and the Oxford Experimental Philosophy Club decide to form a national academy for experimental science.

1661   With the publication of *The Sceptical Chemist*, Boyle establishes chemistry as a separate science to medicine and in 1662, with assistance from Hooke, he discovers the inverse relationship between pressure and volume of gases.

1662   The national academy for experimental science formed by Robert Boyle, John Wallis, Christopher Wren and others receives a royal charter as the Royal Society with Robert Hooke appointed as Curator of Experiments.

1664   After dissecting and investigating brains for more than a decade, Willis publishes *Cerebri anatome*. With illustrations by Wren, who assisted Willis's dissections, it is unsurpassed as a neuroanatomy text for almost 200 years.

1665   Working in isolation over eighteen months from the summer of 1665, Isaac Newton begins his *annus mirabilis*, inventing calculus and the reflecting telescope, devising his laws of motion and a theory of colours involving the splitting and reconstitution of white light through prisms, and formulating his law of universal gravitation. Hooke publishes *Micrographia* and coins the term 'cell' to describe the porous structure of cork. The first edition of *The Philosophical Transactions of the Royal Society*, the first English-language and the world's longest-standing scientific journal, is published.

1666   Following the Great Fire of London, Christopher Wren and Robert Hooke start work on the design and rebuilding of London.

1668   Wallis is first to suggest the law of the conservation of momentum, the first of the conservation laws.

1669   Working in isolation from the scientific establishment in London and now obsessed with chemistry, Newton is appointed Lucasian Professor of Mathematics at Cambridge.

1672   After sending the Royal Society his reflecting telescope, Newton is elected a Fellow and publishes his paper on his splitting of white light into its constituent colours in the *Philosophical Transactions*, leading Hooke to raise the cry of plagiarism that triggered their long-running feud.

1674   In a lecture, Hooke puts forward ideas that more than a decade later he would accuse Newton of appropriating, including laws regarding the motion of bodies and a centripetal force between all bodies akin to gravity.

1675   Appointed first Astronomer Royal, John Flamsteed works on the first great star map of the telescopic age.

1676   Newton writes his now-famous 'standing on the shoulders of Giants' rebuke to Hooke as they put on a public show of patching up their differences; he then retreats altogether from public life to concentrate on alchemy.

1677   Halley is in St Helena, mapping the southern sky.

1678   Hooke determines that the force restoring a spring to equilibrium is proportional to the length of its displacement from equilibrium (Hooke's law).

1679   In a letter to Newton, Hooke suggests an inverse-square law governing gravitational attraction – the basis of the law of universal gravitation that Newton had earlier and independently derived but chosen not to make public.

1682   With Halley's comet in the night sky, Wren, Hooke and Halley meet to discuss laws governing gravitation.

1684   In August, Halley visits Newton in Cambridge, prompting Newton to devise his law of universal gravitation, published in November, and to embark on writing up two decades of scientific and mathematical work in *Principia Mathematica*.

1686   John Ray begins publication of his three-volume encylopaedia of plant life that laid the groundwork for the systematic classification of species.

1687   Isaac Newton publishes *Principia Mathematica*, widely regarded as the greatest scientific work ever written. It states the three universal laws of motion and from them defines the law of universal gravitation.

1698   Thomas Savery's miner's friend is the first practical machine to harness the power of steam and the first patented steam engine, although technically it is an atmospheric engine.

1705   Edmond Halley publishes *A Synopsis of the Astronomy of Comets*, predicting the return in 1758 of the comet previously seen in 1531, 1607 and 1682, now known as Halley's comet.

1712   Thomas Newcomen collaborates with Savery to produce an improved steam atmospheric engine that is the first commercial steam engine and soon becomes the industry standard.

1727   Stephen Hales publishes a book detailing his discoveries of plant physiology, the first measurement of blood pressure, methods of distilling fresh water from sea water, the use of sulphur dioxide as a preservative and the collection of gases over water.

1728   John Harrison builds the first of his series of five clocks whose accuracy enables ships to determine their longitude.

1758   John Dollond, an optician, develops achromatic lenses, thus disproving Newton's contention that chromatic aberration could not be avoided and thereby inventing the achromatic refracting telescope. On Christmas Day, Halley's comet is spotted above present-day Germany, proving his prediction in 1705 that it would return now.

1762   John Hunter publishes his first paper, *The State of the Testis in the Foetus and on the Hernia Congenita*, and goes on to become one of the greatest anatomists of all time and the founder of modern surgery.

1764   Joseph Black discovers the differences between heat intensity (temperature) and heat quantity (a measure of its physical state), and thereby explains the latent heat of substances.

1766   Henry Cavendish discovers hydrogen.

1768   Joseph Banks accompanies Cook as botanist on his first expedition to the South Pacific to verify Halley's prediction of a transit of Venus in 1769.

1769   James Watt patents a more efficient steam engine with a separate condenser.

1772   Daniel Rutherford discovers nitrogen.

1776   Joseph Priestley announces his discovery of oxygen (which he called dephlogisticated air), thereby replacing the four classical elements theory with his own variation of phlogiston theory.

1782   Watt introduces a double-action steam engine with a sun and planet gear to turn the up-and-down motion of the piston into rotary motion. Also used in steam locomotives, it provided the tipping point of the industrial revolution when used to power cotton spinning machines based on Richard Arkwright's water frame, which came out of patent in 1783.

1785   With his publication of *Theory of the Earth*, James Hutton becomes the father of geology.

1790   William Nicholson demonstrates electrolysis, the first evidence that an electric current can bring about a chemical reaction and the reverse of the Italian Count Alessandro Volta's demonstration that a chemical reaction could generate an electric current in a voltaic cell, or battery.

1792   William Murdock creates coal gas and invents the gaslight to illuminate his house, the first building anywhere to be lit by gas.

1796   Edward Jenner infects eight-year-old James Phipps with cowpox then injects an otherwise lethal dose of smallpox to prove the principle of vaccination.

1798   Cavendish determines the weight and density of Earth, thereby calculating Newton's universal constant of gravity.

1800   Humphry Davy discovers laughing gas, the first anaesthetic and the primary hallucinogen of its day.

1802   Having discovered several metals and an amino acid and introduced the concept of chemical weight, William Wollaston is first to observe ultraviolet light and the dark lines on a spectrum, a discovery with profound implications for astronomy.

1803   John Dalton is first to advance a quantitative theory that all matter consists of small indivisible and indestructible particles (which he calls atoms), that these particles are all identical in terms of mass, volume and properties for any one particular element, and that all substances are composed of combinations of these particles of the element, which differ physically from each other only in mass.

1804   The world's first steam railway journey takes place on 21 February, when Richard Trevithick's steam locomotive hauls a train along the tramway of the Penydarren ironworks, near Merthyr Tydfil in south Wales.

1815   With the assistance of Michael Faraday, Davy invents the miner's safety lamp.

1821   Faraday discovers electromagnetic rotation and thereby invents a machine that turns electrical energy into motive force: the electric motor.

1822   Charles Babbage begins work on his Difference Engine, a programmable machine that was the first practical computer.

1823   William Sturgeon invents the electromagnet.

1827   Robert Brown, a protégé of Joseph Banks, discovers the random movement of particles in fluids (now called Brownian motion).

1830   Charles Lyell publishes *Principles of Geology*, a major influence on Darwin, who takes a copy with him the following year on his voyage on HMS *Beagle*.

1831   Faraday discovers electromagnetic induction and thereby invents the dynamo. Charles Darwin sets off on a five-year voyage aboard HMS *Beagle*, which exposes him to observations that lead him to develop his theory of evolution by the survival of the fittest. Robert Brown discovers every cell has a central structure, which he calls the nucleus.

1833   The word 'scientist' is coined by William Whewell.

1838   Brunel launches his SS *Great Western*.

1843   James Joule publishes his discovery of the link between work and heat, leading to the law of conservation of energy, one of the most fundamental laws in science. Brunel launches his SS *Great Britain*, the first iron-hulled screw-propeller-driven ship to cross the Atlantic.

1848   William Thomson (later Lord Kelvin) determines absolute zero.

1854   Alfred Wallace sails to the Far East and observes species differences that lead him to develop his theory of evolution by natural selection of the fittest in parallel to Darwin.

1856   William Perkin, an English schoolboy, discovers synthetic dyes in the form of aniline purple, patents them and initiates the great synthetic-dye industry which leads to the development of synthetic organic chemistry.

1858   The grand theory of evolution by natural selection is presented to the world when Darwin's and Wallace's papers are presented to the Linnaean Society in London, which hardly takes notice.

1859   Darwin publishes *Origin of Species by Means of Natural Selection, or the Preservation of Favoured Races in the Struggle for Life*.

1860   Thomas Huxley becomes the great promoter and populariser of Darwinism against the Church and general resistance. Joseph Swan uses a carbon filament to create light from electricity: the first lightbulb. James Clerk Maxwell establishes that heat is a measure of the average velocity of all the molecules in a substance.

1864   James Clerk Maxwell unifies the three realms of physics – electricity, magnetism and light – in a set of four equations that describe every facet of electromagnetic radiation, immediately making redundant a dozen or more other theories that had been proposed as explanations of light.

1865   Joseph Lister, an English surgeon, develops antiseptics.

1875   William Crookes invents the Crookes tube and studies cathode rays, calling the charged particles he observes a fourth state of matter.

1897   J.J. Thomson discovers the electron, the first subatomic particle.

1899   Ernest Rutherford, a student of J.J. Thomson, discovers alpha and beta particles, although he initially regards them as forms of radiation.

1900   Rutherford discovers gamma rays and that the intensity of radiation emitted from radioactive substances has a half-life.

1903   Charles Barkla discovers that gases scatter X-rays in proportion to their density and molecular weight, leading him to deduce that the more massive the atom, the greater the number of charged particles it contains – the first indication of a connection between the number of electrons in an atom and its position in the periodic table.

1909   Ernest Rutherford discovers that atoms have a nuclear structure and that the charge on the nucleus, rather than the number of electrons orbiting around it, determines the character of the atom; he announces the discovery two years later.

1913   Henry Moseley discovers that the number of protons in the nucleus (the atomic number) determines the position of an element in the periodic table and he predicts four missing elements that are later discovered.

1914   Henry Dale identifies acetylcholine as a neurotransmitter.

1915   William Lawrence Bragg and his father, William Henry Bragg, discover that a crystal of common salt contains no sodium chloride molecules, but a regular geometric distribution of sodium and chloride ions. It has a profound effect on theoretical chemistry and enables them to become the only father–son team to win the Nobel Prize in Physics.

1919   Rutherford splits the atom.

1920   Francis Aston produces mass spectrographs that show most stable elements are mixtures of isotopes.

1925   Francis Aston shows mass numbers of isotopes are not quite whole numbers, the difference representing the nuclear energy that would later be released in devastating fashion when atoms were split to make the change from one element to another in nuclear bombs and reactors.

1927   George Thomson, son of J.J. Thomson, discovers the wave nature of electrons, for which he shares the Nobel Prize with Clinton Davisson. Dirac publishes his equation that provides the definitive statement on the electron, albeit with two answers.

1929   Alexander Fleming announces his discovery of the antibiotic properties of *Penicillium notatum* mould, although his published findings awaken little interest.

1931   After tussling for four years with the implications of his equation, Paul Dirac announces that it predicts the existence of antimatter, a prophecy that is proved by experiment two years later.

1930   Frank Whittle patents his first jet turbine engine.

1932   James Chadwick discovers the neutron. John Cockcroft and Ernest Walton undertake the first efficient nuclear fission reaction by bombarding one atom of lithium with one atom of hydrogen to make two atoms of helium.

1935   Patrick Blackett is first to show the conversion of energy into matter, confirming Einstein's $E=mc^2$.

1936   Alan Turing establishes the concept of a Universal Machine – what is now called a computer program – in his seminal paper, *On Computable Numbers, with an Application to the Entscheidungsproblem*.

1937   Hans Krebs determines the tricarboxylic acid cycle, the fundamental means by which all food materials are metabolised and turned into energy.

1938   Alexander Todd starts work on synthesising all naturally occurring nucleotide components of nucleic acids, opening the door for Wilkins, Watson and Crick to work out the fine detail of DNA structure.

1940   Having discovered a method of extracting and purifying penicillin, Howard Florey and Ernst Chain establish the crucial test that it can protect mice from a lethal injection of *streptococci* bacteria. When over-stretched drug companies refuse to start production, they turn their Oxford lab into a penicillin factory.

1941   Britain's first jet, a Gloster E28 with a single jet turbine designed by Whittle, makes its first test flight.

1942   Harold Jones determines 93,005,000-mile distance of Earth to the Sun to an accuracy of 0.0001 per cent.

1943   With the assistance of Turing, British scientists at Bletchley Park develop Colossus, the world's first programmable digital electronic computer, to help decipher German higher command teleprinter messages.

1945   Dorothy Hodgkin solves the structure of penicillin using X-ray crystallography and a Hollerith punched card calculator.

1947   Cecil Powell discovers the pi-meson, or pion, a subatomic particle that makes up protons and neutrons.

1949   Fred Hoyle discovers nucleosynthesis, an explanation for the creation of carbon and *inter alia* all elements in the universe.

1953   In the 25 April issue of *Nature*, three teams reveal their discovery of the structure of DNA: Francis Crick and James Watson announce their double-helix model without reference to the contribution from Rosalind Franklin's X-ray image, which in a neighbouring paper provides convincing data to support their model. A third paper by Maurice Wilkins suggests a helical structure. It takes another ten years for the model, plucked out of the air to fit the observed data, to be proved experimentally.

1962   John Gurdon announces he has cloned South African clawed frogs, the first cloned animals.

1964   William (Bill) Hamilton publishes his paper on 'the genetical evolution of social behaviour', which lays out the concept of what came to be called the selfish gene and becomes the most cited paper in all science.

1974   Working with the mathematician Roger Penrose, Stephen Hawking discovers that the universe was born in a singularity, which suggests the Big Bang theory is true.

1978   Patrick Steptoe and Robert Edwards report the first birth of an IVF 'test tube' baby.

1984   Alec Jeffreys of Leicester University discovers genetic fingerprinting.

1994   Nicholas Terrett and Peter Ellis, two British scientists working at Pfizer, file a patent for Viagra as a treatment for impotence.

1997   Ian Wilmut reports the birth of the first cloned mammal, Dolly the sheep, born on 5 July the previous year.

2000   John Sulston leads a team at the Sanger Centre in Cambridgeshire that is first to map the human genome.

INDEX

INDEX

Index

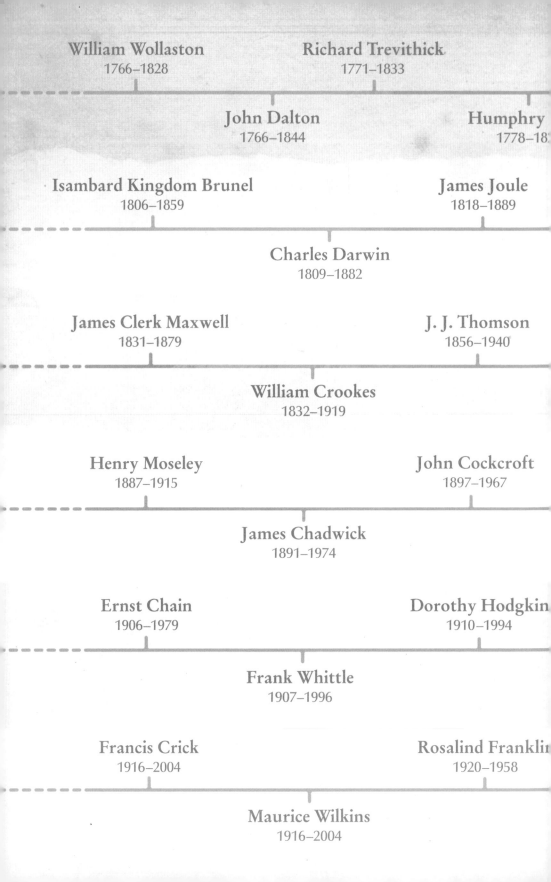